EVERYBODY'S ECOLOGY

EVERYBODY'S ECOLOGY

A Field Guide to Pleasure and Perception in the Out-of-Doors

Clay Schoenfeld

**Original Photographs
by E. William Wollin**

SOUTH BRUNSWICK AND NEW YORK:
A. S. BARNES AND COMPANY
LONDON: THOMAS YOSELOFF LTD

A. S. Barnes and Co., Inc.
Cranbury, New Jersey 08512

Thomas Yoseloff Ltd
108 New Bond Street
London W1Y OQX, England

ISBN 0-498-07850-7
Printed in the United States of America

A/574.5

OTHER BOOKS BY THE AUTHOR
The University and Its Publics
Effective Feature Writing
Publicity Media and Methods
The Shape of Summer Sessions to Come (editor)
Year-Round Education (with Neil Schmitz)
University Extension (with T. J. Shannon)
Wisconsin Sideroads to Somewhere
The American University in Summer (with Donald Zillman)
Cabins, Conservation, and Fun
Canada Goose Management (editor, with Ruth Hine)
Outlines of Environmental Education (editor)

For

SSSS

Let no man jump to the conclusion that he must take his Ph.D. in ecology before he can "see" his country. On the contrary, the Ph.D. may become as callous as an undertaker to the mysteries at which he officiates. Like all real treasures of the mind, perception can be split into infinitely small fractions without losing its quality. The weeds in a city lot convey the same lesson as the redwoods; the farmer may see in his cow-pasture what may not be vouchsafed to the scientist adventuring in the South Seas. Perception, in short, cannot be purchased with either learned degrees or dollars; it grows at home as well as abroad, and he who has a little may use it to as good advantage as he who has much.—ALDO LEOPOLD.

Contents

Preface

A lot of people are discovering ecology these days, but the spirit of ecology isn't really new. In its simplest form it is a recognition of the relevance of nature to human lives, and this recognition has been a hallmark of outdoor buffs for many years. On the other hand, scientists are learning much we didn't know before about man in nature. So this book is a fresh approach to an old phenomenon, using the new language of the ecologist to illuminate what comes instinctively to every outdoor fan—the enjoyment, appreciation, and conservation of natural processes.

This book is a series of sketches about hunting and fishing, hiking and touring, camping and cabins, scenery and sewage, attitudes and action. It is also a primer of ecological principles. In short, it tries to promote perception—perception of our beautiful yet threatened environment, perception of the individual practices and collective policies that can preserve and enhance our natural inheritance.

A salient feature of the book is a set of reports from a squirrel who has gone to the capital as a lobbyist for conservation. Through his eyes we get a comical yet thoughtful view of the oneness of man and environment.

CAS
January, 1971

Acknowledgments

Portions of this book have appeared previously in the following publications and are used with the permission of the respective editors: *Field & Stream, Sports Afield, American Forests, Journal of Soil and Water Conservation, Environmental Education,* (Madison) *Wisconsin State Journal, Wisconsin Sideroads to Somewhere* (DERS, 1966), *Cabins, Conservation and Fun* (Barnes, 1968), *Canada Goose Management* (DERS, 1969), and *Outlines of Environmental Education* (DERS, 1971).

EVERYBODY'S ECOLOGY

Part I

ENVIRONMENTAL ALMANAC

ECOLOGY CAN COME IN FANCY LABORATORY TRAPPINGS, OR IT CAN come in the most ordinary of garbs, because in the final analysis it is simply a way of looking at things, and particularly, a way of looking at the out-of-doors. It is the way of looking at our environment that recognizes the relevance of nature, and the relationship of nature to everybody.

So this is a book about everybody's ecology; in other words, it is about the American out-of-doors—its enjoyment, its appreciation, its conservation—and the ecological rules of the game. It is about citizens in action in environmental management. In short, this book is a guide to pleasure and perception—pleasure in the simple values and rustic charms of outdoor recreation, perception of an awareness of our beautiful yet threatened environment, perception of the principles, policies, and practices that can preserve and enhance our natural inheritance.

In a heaving, tossing, lavender-plumbing world, some may very well give serious question, I suppose, to the relevancy of the out-of-doors. Faced with an "environmental crisis" in every headline, can we continue to practice in good heart such cherished diversions as birdwatching and bluegill fishing? Does the voice of a meadowlark or the tug of a trout impart any message worthy of the hour? Are we simply engaged in a heedless escape from stress? Or is there in fact some significant meaning hidden in a spring hike or a fall hunt?

I believe there are at least three valid reasons for the pursuit of outdoor recreation even in the face of the population-pollution syndrome. One is simply that outdoor recreation is indeed a re-creating experience. We have always felt instinctively that we need "fresh air," literally and figuratively. That the human ani-

15

mal demands an occasional change of pace and change of scene there is increasing scientific evidence. Arnold Toynbee has documented the phenomenon of withdrawal and return that has enriched the course of history. Biochemists have discovered alternating cycles of activity and quiescence in cellular affairs that are to be denied only at the risk of health. Psychiatrists are using outdoor activities to calm and cure their patients. The outdoor fan need make no apologies, then, if he takes to field and stream now and then. He returns a better man, and his family with him.

Exposure to the out-of-doors can take on an inspirational dimension as well. Just as there are no atheists in foxholes, so are there no pessimists on spring hikes or fall hunts. The resurrection of the April earth or the fire-dance of autumn impart a compelling recognition that change is a law of nature.

True, a carpet of shooting-stars, a congregation of warblers, a big moon over Blue Mound, even all of them together speak in a very still, small voice. What indeed is a June sunset in a day of space ships? And yet don't we really need, today as seldom before, something of the serenity and detachment that the out-of-doors has to offer? We can use this intimation of eternity to rout our jitters and clear our vision.

But the real ecological justification for an occasional retreat to woods and waters is the opportunity it provides to acquire a perception of the oneness of our world. To partake of the natural processes by which the land and the living things upon it have achieved their characteristic forms and by which they maintain their existence, to become aware of the incredible intricacies of plant and animal communities, to sense the intrinsic beauty and the creeping degradation of the organism called America.

To do these things is to learn the great lesson: that there are no country problems that are not city problems, no problems of the inner core that are not problems of the open spaces, no local problems that are not world problems, no world problems that are not local problems, no problems of poverty that are not problems of affluence, no problems of crime in the streets that are not problems of conduct in the home; that insects, birds, fish, mammals, water, soil, wilderness, trees, plants, man, Washington, Vietnam, Harlem, and Weyauwega are all part of the same scheme—a sort of intricately woven fabric. Snip one thread and the entire cloth begins to unravel; stitch up one tear and you begin to repair the whole.

Joseph Wood Krutch said it so well:

> We need some contact with the thing we spring from. We need nature at least as a part of the context of our lives. Without nature we renounce an important part of our heritage. On some summer vacation or country weekend we realize that what we are experiencing is more than merely a relief from the pressures of city life: that we have not merely escaped from something but also into something; that we have joined the greatest of all communities, which is not that of man alone but of everything which shares with us the great adventure of being alive.

In short, there is a life of the woods and the fields and the streams that can *mean* something. Our lives will be dull and unrewarding unless we come to appreciate the solitude and the significance of natural surroundings.

This does not mean we become hermits. Hermits are distinguished more by their fear of mankind than by their love of nature. Nor am I suggesting escape for escape's sake. But I am suggesting the value of an escape from wordiness—as a healthy resistance to the false notion that everything of significance can be heard in speech or read from the printed page.

Beyond words there is a life experience that words may evoke but for which they themselves are a poor substitute. In this sense outdoor recreation, away from closed-in spaces and away from crowds, may offer unique opportunities for personal growth. Away from the clutter of words, there is what the physiologists call the wisdom of the body. It is older than any culture. Its controlling mechanisms run deep into every fiber. You swim or you sink.

Most of us listen too much to others and to ourselves. At times we use words to the detriment of understanding. To seek the meanings that lie beneath the rhetoric we should go back, whenever we can, to the world of nature, to hear in the stillness of the woods what Mr. Wordsworth called "the sad music of humanity."

That is why outdoor life stirs people so. It is a healthy return—

if only for a brief time and under simulated conditions—to a life
the whole race of man once knew. It is Mr. Thoreau cutting a
broad swath and shaving close. It is Mr. Frost going out to clean
the pasture spring.

Of course it is not enough to sit like a pumpkin by some local
Walden Pond, or to walk unknowingly through the meadow under a scarcely noticed pattern of stars. There are wonderfully
meaningful forms in nature, from the microscopic to the galactic,
and the way to sense them is to combine reading with direct
observation. It is a matter of studying and searching and com-
paring, and at the same time enjoying the outdoor life on a
plane of personal experience.

To do this it is not necessary to become a scientist, but it
is good to know something of their findings. Why are there six
frogs and not ten in your pond? Why is there only one pair of
bluejays around your cabin? It is a fascinating pursuit to go
from simple observation to a little study and then back to more
observations. This is the meaning of ecology for everybody.

I am not so naive as to suggest that every lathe operator with
a spinning rod is a potential Thoreau, that every member of the
Audubon Society is an Aldo Leopold in disguise, or that every
delegate to a state conservation congress is a homespun Albert
Schweitzer. But I do proceed here on the assumption that exposure to the out-of-doors, however casual or however intense,
directly or vicariously, can be a doorway to the ecological understanding of our utter interdependence with our environment
and with life everywhere, and to the development of a culture
that will secure the future of an environment fit for life and fit
for living.

So come along, then, as we travel down the sideroads of America in search of those pleasures in outdoor recreation that are
everybody's for the finding, as we discover some ecological
ABCs, and as we talk about conserving those resources that are
inextricably linked not only to economic prosperity but to the
inner prosperity of the human spirit.

1. The ABCDs of Ecology

"A" Stands for Approaches

THE WORD ECOLOGY HAS BEEN LYING IN THE DICTIONARY FOR SOME 100 years, but it has been only recently that the term has begun to flit about in everybody's vocabulary, like a bat emerging suddenly from a cave into bright sunlight.

Just exactly what does the word mean? What *is* ecology? What does it *do?* What can *I* do with *it?* Is it one of those complex things that only a scientist can really understand and employ, or is it something within the ken of the average person?

There are all sorts of approaches to ecology, because ecology is a very rich word; that is, we can put it to many uses. Like the celebrated committee of blind men examining an elephant, we can come up with all sorts of answers to our questions.

From one point of view, we can define ecology scientifically as "the study of the interrelations of organisms and their environments." This is the classic way the term is used in college catalogues, to distinguish it from other basic divisions of biology, such as physiology or embryology.

In this regard, it may be helpful for us to note that ecology is a word we derived from the Greek root "oikos," meaning "house." So ecology is concerned literally with "houses," or more broadly, "surroundings," as opposed to cells or organs or individual organisms themselves. Perhaps the best meaning for this sense of the term is: "the science of community."

Viewed a little more broadly, ecology is not simply the study *of* something; it *is* that thing. In other words, the word refers

19

in a grander sense to the makeup and operations of the living
world, to "the structure and function of nature," including man
himself.

We can use the term to refer to a property of a *piece* of nature, like a lake or a covey of quail. By this we mean the whole
complex of living conditions of a body of water or a group of
birds. So we can speak of "the ecology" of a park, or a walleyed
pike, or so on.

It used to be that we applied this ecology concept only to
living organisms in "nature." In more recent years we use the
term in connection with *any* structure or organization. So we
can speak of the ecology of a suburb, or of a labor union, or of
the conservation movement itself—meaning, still, the *relationship* among an "organism" and its environments.

And thus we come to the more philosophical meaning of ecology. In its broadest sense it is *a way of looking at things*—a
viewpoint that sees not the things themselves so much as their
connections with other things. So it is a concern with processes,
with the myriad of mechanisms that make up the web of life
of whatever we are looking at. In other words, in an ecological
look at an oak woodlot, for example, we concentrate on the
spaces *between* the trees, spaces that are actually filled with all
sorts of mechanical and chemical operations, or processes.

What is more, in this eyeball shift from *thing* to *process*, we
recognize that the human observer himself is an integral part
of the picture. So ecology really becomes a way of looking at
"things" from the inside out, rather than from the outside in.

Who, then, is an ecologist? Well, obviously, a modern ecologist can be a scientist, although not all scientists are ecologists
by any means, not even all those who carry "ecology cards."
Some "ecological" pundits get so concerned with bits and pieces
of nature that they literally do not see the forest for the trees.
Other professors of ecology, on the other hand, are today's leading representatives of ecological thinking. But so are some politicians and some English professors—those who see that everything is connected to everything else.

Prehistoric man was a superb ecologist. He could not run fast
enough to escape his enemies; if caught, his teeth and claws

were small protection. So he had to become a student of his relationship to the veldt. Peering from his hiding place in the bushes around a clearing or from the opening of his cave, his science was the practical kind. His laboratory was the place he lived. The success of his observations could be measured by the fact that he made it through the night—or didn't.

Today an ecologist can be simply anybody who appreciates his two-way-street relations with his environment and with all of life everywhere. He is the hunter who knows that you look for whitetails not on a prairie but along the edge of a forest. He is the fisherman who understands why DDT sprayed on a carrot patch can affect the eating quality of lake trout caught miles away. She is the housewife who sees the connection between her automobile engine and air pollution.

We give this sense of interlocking community the term *ecological awareness*—an awareness that the community to which each of us belongs includes soil, water, plants, animals, and people. Then, if we have an *ecological conscience* as well as ecological awareness, we know that an action is *right* only if it tends to protect the "health" of our man-land community—that is, its integrity, stability, and beauty—and we act accordingly, regardless of what may be momentarily convenient or profitable.

So what we used to call "conservation" becomes *applied ecology*—the ethics that should govern the relations between humans and the living landscape, the "resistance movement" that challenges anybody's right to pollute or over-populate the environment.

In a nutshell, then, while there are a number of approaches to the use of the term *ecology*, when we "take all the feathers off of it," as the saying goes, ecology is a way of looking at our world that says to us: "I am a part of my environment, and my environment is a part of me."

"B" Stands for By-Laws

Now that we have seen the various "ecological" approaches that are ours to take today, let's back up a bit and say, "All right, suppose we concentrate for a moment on a study of the operation of nature. What, then, are the by-laws, so to speak, that we will discover?"

To answer that question, we first have to assimilate a basic ecological frame of reference, the idea of an *ecosystem*. *Eco*, as we have discovered, is shorthand for the idea of an organism and its environment interacting. *System* implies a regular, orderly way of doing something. So an ecosystem is a set or arrangement of things in the natural world so related or connected as to form a unity of an organic whole.

Ecosystems come in all shapes and sizes. A forest is an ecosystem, but so is the world of the mites living in the fallen needles on the forest floor. The aquarium in your livingroom is an ecosystem. So, of course, is a pond.

A pond is a particularly good example of an ecosystem, because it exhibits a recognizable unity both in structure and function. Its ecology includes the food-chain, with which we are familiar from high-school biology—plants converting sunlight to food, tiny water fleas grazing on tiny plants, minnows eating the water fleas, a bass eating a minnow, a man catching the bass. Yes, man is a part of the pond ecosystem; man drinking, swimming, fishing, plowing the slopes of the watershed, spraying the shallows, drawing a picture of the pond, formulating theories about the world based on what he sees in the pond. He and all the other organisms at and in the pond act upon one another, engage the earth and the atmosphere, and are linked to other places by a network of processes like the threads of protoplasm connecting cells in living tissues.

In the largest sense, then, the world is an ecosystem, and man is but one of many passengers, all of them related to each other and to their environment. This super-ecosystem we call the biosphere.

Regardless of size or shape, every ecosystem has four basic "parts." First, there are *basic elements* and compounds of the environment, like oxygen and iron. Second, there are *producers*, the green plants that have the ability to turn basic elements into food. Third, there are *consumers*, animals that feed on plants or on the other animals or both. Finally, there are *de-*

composers, bacteria that break down dead plants and animals into the basic elements again.

Connecting these parts or "citizens" of an ecosystem are three basic processes: energy flow, nutrient cycles, and information exchange. These processes are really what an ecosystem is all about. They are what convert a collection of parts into a "machine." A brief examination of each process will reinforce our appreciation of the utter interrelationship of organism and environment.

Energy flow is "where it's at," as the saying goes. Radiant energy, in the form of sunlight, is captured by green plants and turned into chemical energy in the form of carbohydrates. This chemical energy is then transferred through various levels of consumers. Energy is the "gas" that makes the ecoengine go. Man, mouse, or microbe, we get our energy from the sun via green plants. There isn't any other flow. And it's a one-way street.

There are three types of *nutrient cycles.* The first is a liquid cycle. Water falls on the earth from the atmosphere in the form of rain. It goes back by evaporation as well as by transpiration, the loss factor from green plants, meanwhile providing essential ingredients for ecosystem citizens and their life styles. Another cycle is a gaseous cycle; carbon is a good example. It moves from the atmospheric reservoir through producers to consumers and from both these groups to decomposers, and thence back to the reservoir. A third type of cycle is a sedimentary—or mineral —cycle, involving such nutrients as phosphorous. All told, there are some 20 basic nutrients that cycle through an ecosystem and its citizens in one way or another, linking organism and environment in a web of life.

Besides a flow of energy and a cycling of nutrients, an ecosystem exhibits a *flow of information.* Birds start to sing in the morning when a certain light intensity imparts a message. Rodents over-crowded in a cage get a message that compels them to stop breeding. The seeds of plants do not sprout until temperature gives them the word. Man reads the skies and plans a picnic. His appearance in a park will be a signal to ants and mosquitoes. And so it goes. The exchange of information in the

simplest of ecosystems is at least as complicated as watching a
splashdown via TV satellite.

An ecosystem, then, is a community made up of all the organisms in a given area, interacting with themselves and with
the physical environment via a flow of energy, a cycling of materials, and an exchange of messages. The environment shapes
the community, the community modifies the environment, and
each plays upon the various citizens of the ecosystem.

Each ecosystem citizen occupies a certain *niche*. That is, he
participates in the structure and function of the system in a
particular way; he practices a specific profession, as it were. A
niche can be occupied by only one species at a time. The more
niches and species in the ecosystem—the more diversified it is,
as we say—the more stable it is; that is, the less susceptible it
is to collapse if a strand in the web is severed for one reason
or another. Any ecosystem will have a comparatively small number of "common" species making up the bulk of the community, and a large number of "rare" species. Yet a single rare
species can be as important to the operation of the system as
can a common species, just as a car won't work without a car-
buretor even though it has only one.

The purpose or goal of an ecosystem is to survive and flourish
in a healthy, steady state. To do so it is aided by two opera-
tional by-laws that work against each other: the lust of a population for life, the drive to reproduce, versus a variety of natural
controls on populations (a population being an assemblage of
the same species of plant or animal).

One mechanism regulating populations is a shortage of a basic
ingredient, like sunlight, water, or nitrogen. Another mechanism
is the "two's company, three's a crowd" factor, in which sheer
lack of living space inhibits reproduction. A third mechanism
is predation, in which one species literally eats away another.
It is in the competition for life-support factors that species adapt,
evolve, and survive, and by putting a ceiling on that survival
ecosystems make their contribution to a balanced nature.

Ecosystems don't live forever. That pond we spoke about will
eventually become a bog and then dry land. An abandoned corn
field becomes a weed patch, then a brush patch, and finally a
forest. We call this process ecological *succession*. The various

stages in succession have different characteristics. A "young" stage, like a well-fertilized pond, is very productive, but an "old" stage, like a maple forest, is more stable. Many species, including man, seem to survive best in the "edge" ecosystems between communities in different stages of succession.

Man affects ecosystems in many ways. Sometimes he aids nature. More often he disrupts natural structures and processes. That disruption in turn can threaten his own survival. This is the biggest ecological by-law of all.

"C" Stands for Concepts

If ecology is a special way of looking at things, and if what we look at is an ecosystem, small or large, what are some of the basic concepts or understandings that we glean? Particularly, what are the ecological rules of the game that can be applied to our own lives?

Some of the following ecological concepts are rather profound. Others are more picayune. Some are fundamental. Others say about the same thing in different ways. Taken together, however, they pretty well sum up what we need to know about man-environment relationships in our search for survival.

When the bell tolls, it tolls for me. I am a part of my environment and my environment is a part of me. When I maim the earth, I wound *me*. When I pollute a stream, I poison *me*. When I fill the sky with smog, I choke *me*. For this is my world, I am the world and the world is *me*. This is the essence of ecological understanding.

Everything is connected to everything else. "This," said ecologist Barry Commoner recently, "is the first concept of ecology." John Muir said it long ago: "When we try to pick out anything by itself, we find it hitched to everything else in the universe." The ecologist Aldo Leopold put it poetically: "When you pick a pasque flower, you disturb a star." However the concept is phrased, it is at the heart of ecological understanding. The continuing functioning of any organism depends on the interlinked function of many other organisms. As writer William Bowen has emphasized, the seemingly autonomous oak in the forest depends upon microscopic organisms to break down fallen leaves, releasing nutrients that can be absorbed by its roots. Seemingly autonomous man ultimately depends on plant photosynthesis for his food.

Everything's got to go somewhere. This is Commoner's second great concept in ecology. It is simply a popular rephrasing of the well-known law of physics: that matter can be changed from one form to another, but it cannot be destroyed. This is the law of nature that causes pollution and complicates pollution abatement. Pollution of air, water, and soil occurs when the load of wastes discharged into them exceeds the capacity of the natural systems for absorption and regeneration. These wastes are the leftovers of technology—the unsavory by-products of extraction, production, and advanced mass consumption. The complication of pollution management is that matter is indestructible. By treatment, its form can be changed from solid to liquid to gas. Or energy can be produced, or converted to noise. But "residuals" are still there in one form or another, to be discharged into the environment where they can pollute the air, the water, or the soil. For example, a mass of solid wastes can be reduced by burning, but in the process the air is polluted, often with toxic substances. Stack gases can be washed to remove dust and dirt and some noxious gases, but these are carried away in the wash water, to pollute streams and lakes. The problem can be alleviated through increasing the efficiency of each extraction, production, and consumption process so that the proportion of useful product at each stage is increased, and residuals are decreased. Residuals can also be reduced by recycling; that is, by reusing wastes wherever possible in the production and consumption system, in place of some of the new materials that would otherwise be used.

There's no such thing as a free lunch. We pay for everything we do, in some way or another, at some time or other. In ecosystems causes and effects are often widely separated. Accordingly our interventions often yield unexpected consequences. As William Bowen points out, after years of spraying persistent pesticides to kill insects, we find that we have come close to wiping out a national symbol, the bald eagle: concentrated through food chains, pesticides accumulate in the tissues of eagles and certain other birds to the point of impairing reproduction. We drain Florida swamplands and learn later on that by reducing the outflow of fresh water into estuaries we have increased their salinity and thereby damaged valuable breeding environments for fish and shrimp. The Aswan Dam impounds silt that would otherwise be carried downstream, so the Nile no longer performs as richly as before its ancient function of renewing fields along its banks. The fertility of the Nile Valley is therefore declining. That is only one variety of ecological backlash from this triumph of engineering. With the flow of the river reduced, salt water is backing into the Nile delta, harming farmlands there. And in time, some authorities predict, the flow of Nile water to new farmlands through irrigation canals will bring on a calamitous spread of schistosomiasis, a liver disease produced by parasites that spend part of their life cycle in the bodies of snails. So, as the philosopher said, nothing exists for whose nature some effect does not follow.

Trees do not grow to the sky. This saying, says William Bowen, expresses another basic theme of ecology. Nothing grows indefinitely—no organism, no species. Much more commonly than many might suppose, animal species limit their own growth: rates of reproduction respond to crowding or other signals so that total numbers remain commensurate with the resources of the ecosystem. In the over-all ecosystem of the earth, total animal energy is limited by the amount of solar energy plants convert into organic compounds. Since combustion consumes oxygen, the amount of combustion the earth can sustain is limited—other limiting factors aside—by the ecosystem's net production of free oxygen. Ecologist Ed Kormondy cites this example of the limitation concept:

In 1899, shortly after the English sparrow was introduced into the United States, it was estimated that in ten years a single pair of sparrows could give rise to 275,716,983,698 descendants, and that by 1916-1920 there would be about 575 birds per 100 acres. By the 1916-1920 period, however, there were only 18 to 26 birds per 100 acres, less than five percent of the expected. There were, then, forces acting against this geometric increase of the sparrow; similarly, forces must be operating against the increase potential in houseflies and elephants and other organisms as well. It is apparent that there are two opposing forces operating in the growth and development of a population; one of these is inherent in and characteristic of each species population—the ability to reproduce at a given rate. Opposing this is an inherent capacity for death.

We can never do merely one thing. When we intervene in a complex system so as to produce a certain desired effect, we always get in addition some other effect or effects, usually not desired. As ecologist Garrett Hardin has said: "Systems analysis points out in the clearest way the virtual irrelevance of good intentions in determining the consequences of altering a system." Ecologists are accustomed to looking at nature as a system, and if we had paid more attention to them we might have been spared a lot of backlash. In trying to reduce insect damage to crops, for example, we might have made more use of specific biological or biochemical means of control and less use of persistent broad-spectrum insecticides. We might now, accordingly, have more birds in our countryside and less DDT in our streams —and in some places, fewer harmful insects in our fields, William Bowen points out.

The "one-problem, one-solution" approach won't work. Many essential life-cycle resources, not to mention recreational and esthetic needs, are best provided man by the less "productive" landscapes. In other words, the landscape is not just a supply depot but is also the *oikos*—the home—in which we must live. Until recently mankind has more or less taken for granted the gas-exchange, water-purification, nutrient-cycling, and other protective functions of self-maintaining ecosystems, chiefly because neither his numbers nor his environmental manipulations have been great enough to affect regional and global balances. Now, of course, it is painfully evident that such balances are being affected, often detrimentally. Ecologist Eugene Odum emphasizes that the "one problem, one solution approach" is no longer adequate and must be replaced by some form of ecosystem analysis that considers man as a part of, not apart from, the environment. Society needs, and must find as quickly as possible, a way to deal with the landscape as a whole, so that technology will not run too far ahead of our understanding of the impact of change.

The whole is greater than the sum of the parts. In nature there aren't two distinct compartments—organisms and environment. For any organism, other organisms constitute part of the environment. And the physical environment itself is, in part, created and maintained by organisms. Atmospheric oxygen, necessary to the survival of life on earth, is itself a product of life. A hardwood forest creates its own peculiar environment in which seedlings of only certain plant species can grow to maturity. The recognition that organisms and their physical environment are interacting parts of a system, and that this ecosystem has a dynamism of its own—this is the heart of ecology. Ed Kormondy supplies this example of the point: Energy flow, nutrient cycling, population self-regulation—these are the significant properties of ecosystems. None of these ecological processes occurs in isolation; each is manifested by particular assemblages of different species populations in particular environments. Thus, the flow of energy, the cycling of nutrients, and the regulation of populations occur in an assemblage that may consist of broomsedge, field mice, and weasel populations, while a few feet distant the assemblage may consist of populations of herbs, rabbits, and foxes—ecological communities.

The earth is a "closed-system" spaceship. Eugene Odum says a good way to visualize an ecosystem is to think about space travel. For a short journey, such as a few orbits around the earth, man does not need to take along a self-sustaining ecosystem since sufficient oxygen and food can be stored in the capsule to last for a short time. For a long journey involving a number of astronauts, such as an expedition to one of the planets, man must engineer himself into a more independent ecosystem. Such a self-contained spacecraft must include all four of the basic components we have discussed (producers, consumers, decomposers, and nutrient substances) in such proportion and diversity as to maintain a stable environment capable of adjusting to the incoming solar radiation as do the earth's ecosystems. A small capsule with a few components might function outside the biosphere for a short time, but a larger, more diverse system would be more stable and safer for a longer time—if we are to judge from what we observe on earth. As far as that goes, the earth itself can be thought of as a spaceship. As writer Laurence Pringle points out, when an astronaut lands on the moon and looks at the earth, he might think, "The earth is a spaceship, too. It is on a long trip through space, and all life on earth depends on the supplies aboard." Unlike most spaceships, the earth receives some supplies from outside—mainly energy in the form of sunlight. Green plants need sunlight in order to make food, and all life on earth—from earth worms to elephants—depends on green plants for food. Without sunlight, life would end. There seems to be no danger of spaceship earth losing its daily shipments of energy from the sun. But even with such a continuous supply, life aboard spaceship earth could come to an end. As scientists look into the future, they worry about the other supplies aboard: soil, minerals, water, air, plants, wildlife. These are the natural resources of the biosphere, the earth's main supplies. They are limited. The growing numbers of people on earth must soon learn to take better care of the limited supplies aboard their spaceship. There is no choice. After all, it is the only earth we have, and it is a closed system.

For want of a wheel, the clock stops. Author Don Fabun asks us to think of an ecosystem as resembling the inner workings of a clock. Radiant energy from the sun is the force that winds the mainspring. The energy is then distributed throughout the works. There is no such thing as a non-essential wheel. Each absorbs energy, uses a little, and passes what is left on to the next. Should one little wheel fail, the clock stops. So it is in nature. We may not know or understand exactly what role each wheel plays, but we can be sure it has one. From one perspective, man is no more important than any other wheel. From another point of view, he is. He seems bent on disrupting the workings of the clock. He may oversimplify the system, he may overload it, he may drench it with new chemical compounds that alter the timing, he may transfer wastes from one system to another. Eugene Odum points out that man, of course, has also had a considerable influence on the makeup of many ecosystems, since he frequently removes or introduces species. We might think of this effect as a sort of ecosystem surgery. Sometimes the surgery is planned, but too often it is accidental. Where the alteration involves the replacement of one species with another in the same niche, the over-all effect on the function may not be great. However, in many cases severe imbalances have resulted, often to the detriment of man. How to predict better the results of ecosystem surgery, and thereby intelligently prescribe or avoid the removal of vital parts or the addition of cancerous parts, is one of the major objectives of ecological research. For example, although mussels are a relatively minor component of a marsh in terms of energy flow, they have proved to have a major effect on the cycling and retention of valuable phosphorus. The mussel is not particularly important as a direct source of food for man or other animals, but the species is important as an agent helping to maintain fertility. Here is an excellent illustration of the point that species in nature may have great value to man in an indirect way not apparent on superficial examination. A species does not have to be a link in man's food chain to be valuable. Man needs the help of many species in maintaining his environment.

A chain is no stronger than its weakest link. As Eugene Odum writes, any factor that tends to slow down potential growth in an ecosystem is said to be a *limiting factor*. The idea that organisms may be controlled by the weakest link in the ecological chain of requirements goes back at least to the time of Justus Liebig, who was a pioneer in the study of inorganic chemical fertilizers in agriculture. Liebig was impressed with the fact that crop plants were often limited by whatever essential element was in short supply, regardless of whether the total amount required was large or small. Liebig's "law of the minimum" has come to mean that the rate of growth is dependent on the nutrient or other conditions present in the minimum quantity in terms of needs and availability. If we extend this idea to include the limiting effect of the maximum (that is, too much can also limit) and recognize that factors interact (that is, short supply of one thing affects requirements for another thing not in itself limiting), we end up with a working principle that is very useful in the study of any specific ecosystem. We may restate the extended concept of limiting factors as follows: the success of a population or community depends on a complex of conditions; any condition that approaches or exceeds the limit of tolerance for the organism or group in question may be said to be a limiting factor. Although the incoming energy of the sun and the laws of thermodynamics set the ultimate limits in all of the biosphere, different ecosystems have different combinations of factors that may put further limitations on biological structure and function. The chief value of the limiting factor concept lies in the fact that it gives the ecologist an "entering wedge" into the study of complicated situations. Environmental relations are indeed complex, so it is fortunate that not all factors are of equal ecological importance. Oxygen, for example, is a physiological necessity to all animals, but it becomes a limiting factor from the ecological standpoint only in certain environments. If fish are dying in a polluted stream, for example, oxygen concentration in the water would be one of the first things we would investigate, since oxygen in water is variable, easily depleted, and often in short supply. If small mammals are dying in a field, however, we would look for some other cause, since oxygen in the air is constant and abundant in terms of need by the population and, therefore, not likely to be limiting.

Variety is the spice of life. If you find an empty lot, mark off a square yard, stretch a gauze net over it, and then remove all the plants, insects, and other animals in that little square, you will discover at least two things. First, there is a great amount of life going on there. Second, if you divide your harvest up into species, you will find that while there is one species that outnumbers the others, there are numerous other species. This is so because nature stocks each square of earth with a sufficient variety of genes so that if there should be a change in the environment some of the "minority groups" will probably survive. Now if you try the same experiment on a square yard of cultivated land, like a lawn or a cornfield, you'll find very little but dominant species. Such man-made environments are highly vulnerable to environmental change, Don Fabun emphasizes. In short, as Eugene Odum says, the advantage of a diversity of species—that is, the survival value to the community—lies in increased stability. The more species present, the greater the possibilities for adaptation to changing conditions, whether these be short-term or long-term changes in climate or other factors. Or to put it another way, the greater the gene pool the greater the adaptation potential. Occurrences in certain areas of Long Island Sound, once a prolific source of oysters, are an example of an adaptation that would not have been possible if rare species had not been present. The development of large-scale domestic duck farming on shore introduced large amounts of organic manure into the shallow waters. The dominant or abundant phytoplankton producers of these waters were unable to tolerate the changed conditions, but several other species that had formerly been very rare were able to tolerate and exploit the organic materials, and soon became very abundant. The productivity of the ecosystem was thus maintained because producers were present that could "take over." (Unfortunately, in this case, the oysters could not use the new phytoplankton as food and the oyster industry was depressed by the duck industry.)

Cut the root and the plant dies. A pebble, a drop of water, a blade of grass—bathed in the radiance of a beneficent star: these are the magic ingredients whose constant interaction is the foundation of that fragile film we call the biosphere, writer Don Fabun emphasizes. These are *all* the tools we have; to destroy or abuse them, to interrupt their function, is to destroy ourselves. All the acts of parliaments and potentates can't alter this fact of life.

You can put only so many sardines in a can. Any animal—man, mouse, or amoeba—needs a certain amount of a given quality of range to exist in health and beget its kind. A given set of conditions, says Durward Allen, produces not only certain species but also given numbers of individuals. Populations, in other words, seek their own levels. We call this the *carrying-capacity* concept. It is well known in wildlife management, where we have discovered, for example, that there are strict and measurable limits to the number of quail a back 40 will support, or the amount of trout a pond will hold, no matter how many you stock. That a certain area of land or water will likewise support only so many humans, at least in terms of psychic health, there is increasing evidence. Try to crowd too many people into a park, for instance, and some will turn and leave, because you have exceeded the capacity of the park to provide them the solace they seek.

We are a weed species. A weed, says Webster, is "an undesired, uncultivated plant that grows in profusion so as to crowd out a desired crop, disfigure a landscape, and so on." This is a pretty apt description of the human animal. Our mushrooming population, our rampant technology, and our fragile biosphere are on a collision course. We act like weeds in four basic ways, says biologist Clifford Humphrey: (1) We may be throwing the world's energy budget out of balance by discharging more heat into the atmosphere than it can handle. (2) We interrupt natural nutrient cycles; for example, we take many crops from field, forest, and sea without returning any organic materials to whence they came. (3) We contaminate our household with all manner of pollutants. (4) We destroy the surface of our world with axe, bulldozer, dredge, and cement. Unless we become constructive citizens of our ecosystem, our fate as a weed species is predictable: we will ultimately destroy our life-support system.

Population control is a natural phenomenon. At a critical time in the life history of a given population, a physical factor such as light or a nutrient may be significant as a regulatory agent; at another time, parasitism, predation, or competition, or even some other physical factor, may become the operative factor. As complex and as variable as the niche of any species is, it is unlikely that this regulation comes about by any single agency. However, there does appear to be considerable and mounting evidence to suggest that populations are self-regulating through automatic feedback mechanism. Various mechanisms and interactions appear to operate both in providing the information and responding to it. With the exceptional case of a catastrophe, the stimulus to do so appears to depend directly on the density of the population. Kormondy concludes: the end effect is one of avoiding destruction of a population's own environment and thereby avoiding its own extinction.

There is such a thing as "too much of a good thing." The most pleasant and certainly the safest landscape to live in is one containing a variety of crops, forests, lakes, streams, roadsides, marshes, seashores, and "waste places"—in other words, a mixture of communities of different ecological ages. As individuals we more or less instinctively surround our houses with trees, shrubs, and grass, at the same time that we strive to coax extra bushels from our cornfield. We all consider the cornfield a "good thing," of course, but most of us would not want to live there, and it would certainly be suicidal to cover the whole land area of the biosphere with cornfields, since the boom and bust swing in such a situation would be severe. The basic problem facing organized society is determining in some objective manner when we are getting "too much of a good thing." This is a completely new challenge to mankind because, up until now, he has had to be concerned largely with too little rather than too much. Thus, concrete is a "good thing," but not if half the world is covered with it. Insecticides are "good things," but not when used, as they now are, in an indiscriminate and wholesale manner. Likewise, water impoundments have proved to be very useful man-made additions to the landscape, but obviously we don't want the whole country inundated. Vast man-made lakes solve some problems, at least temporarily, but yield comparatively little food or fiber, and because of high evaporative losses, they may not even be the best device for storing water; it might better be stored in the watershed, or underground in aquafers, as Eugene Odum suggests. Another example: phosphorus is a key element in the ecology of a lake, but too much phosphorus produces smelly scum—too much of a good thing.

One organism's poison is another's treat. The presence of large quantities of hydrogen sulfide occurring in the deeper portions of aquatic ecosystems is inimical to most life; its presence probably accounts for the absence of higher animals below 200 meters in the Black Sea, and has been considered responsible for killing fish in valley impoundments polluted by pulp-mill effluents, which are rich in sulfates. Yet some colorless sulfur bacteria oxidize hydrogen sulfide to elemental sulfur, and other species oxidize it to sulfate; other species oxidize sulfide to sulfur, and still others oxidize sulfur to sulfate. For some species, even those of the same genus, the oxidation processes can occur only in the presence of oxygen; for others, oxygen availability is irrelevant. In like manner, the mercury with which a farmer cures his wheat seeds can show up as poison in the flesh of pheasants—and men.

Nature abhors a vacuum. Ed Kormondy points out that, although Baruch Spinoza, seventeenth-century Dutch philosopher, was not an ecologist, his aphorism "Nature abhors a vacuum" quite aptly describes a major ecological phenomenon. Bare ground, either on land or under water, seldom remains that way for very long. "Nature" in the form of vegetation moves in with dispatch. As characteristic as the colonization process itself is the subsequent series of sequential replacements that occur on the site. The structural changes in an open field no longer cultivated, or in an unmanaged farm pond, are both marked and easily recognized; such structural changes have been well studied by ecologists. As a result, the sequence and timing of communities on given sites and in given regions can be predicted with considerable reliability. Thus, in the deciduous forest region of Indiana, ecologists Frederick Clements and Victor Shelford showed that the different communities initially present in such diverse habitats as flood plains, sand ridges, shallow and deep ponds, and clay banks all demonstrated a quite predictable series of changes and that each culminated in a stable ecosystem—a beech-maple forest.

Ecology is more complex than we can think. When he looks closely at any ecosystem, the ecologist invariably comes upon complexity, an intricate web of interrelations. A diagram showing the movement of a single chemical element through an ecosystem can get exceedingly complicated. In the ecosystem of man, which includes institutions and artifacts that themselves impinge upon and alter the environment, the interrelations are unimaginably complex. This great web, the ecologist Garrett Hardin has said, "is not only more complex than we think. It is more complex than we *can* think." We would all benefit greatly, writer William Bowen says, by borrowing from the ecologists their willingness to accept and try to puzzle out complexity, and their habit of sustained, open-eyed observation of what actually goes on.

To appraise the nature of ecosystem complexity, we may review the work of psychologist V. A. Graicumas, who calculated the growth of problems faced by a supervisor as assistants with related work were added to his responsibilities. Deriving an appropriate formula, Graicunas solved for the increasing relationships as follows:

Number of assistants or functions	Number of possible relationships
1	1
2	6
3	18
4	44
5	100
6	222

7	490
8	1,080
9	2,376
10	5,210

We need go no further than ten in the series, says ecologist Durward Allen, since it illustrates beyond question that the addition of individuals or functions in this relatively simple organization gives rise to an exponential increase in relationships. At the root of the problem lies the significant fundamental difference in the *rate of growth* between arithmetical progression, which grows by addition, for example, 2, 4, 6, 8, 10, etc., and geometrical progression, which grows by multiplication, for example, 2, 4, 8, 16, 32, etc. Even the simplest ecosystem thus becomes very complex.

Ecology is "subversive." Bowen emphasizes that the recurrent themes of ecology run counter to some old ways of perceiving and thinking that are deeply ingrained in the prevalent world view of man. We believe in limitless growth (or did until recently); ecology tells us all growth is limited. We speak (or spoke until recently) of man's "conquest" of nature; ecology tells us we are dependent for our well-being and even survival upon systems in which nature obeys not our rules but its own. Our scientists and engineers, and our social scientists too, proceed by isolating and simplifying; ecology tells us to heed existent complexity and patiently try to trace out its strands. In a sense, then, ecology is subversive. A few years ago, ecologist Paul B. Sears called it "a subversive subject," and the editors of a recent compilation of essays on the ecology of man entitled their book *The Subversive Science.* In that book ecologist Paul Shepard says the ideological status of ecology is that of a *resistance movement.* Its Rachel Carsons and Aldo Leopolds are subversive. They challenge the public or private right to pollute the environment, to systematically destroy predatory animals, to spread chemical pesticides indiscriminately, to meddle chemically with food and water, to appropriate without hindrance space and surface for technological and military ends. They oppose the uninhibited growth of human populations, some forms of "aid" to "underdeveloped" people, the needless addition of radioactivity to the landscape, the extinction of species of plants and animals, the domestication of all wild places, large-scale manipulation of the atmosphere or the sea, and most other purely engineering solutions to problems of and intrusions into the organic world. If naturalists seem always to be *against* something it is because they feel a responsibility to share their understanding, and their opposition constitutes a defense of the natural systems to which man is committed as an organic being.

Man is a plain member and citizen of the land-community, not its conqueror. So the practice of conservation must spring from a conviction of what is ethically and aesthetically right, as well as what is economically expedient. A thing is right only when it tends to preserve the integrity, stability, and beauty of the community, which includes the soil, waters, fauna, and flora, as well as people. The direction is clear, and the first step is to throw your weight around on matters of right and wrong in land use. Cease being intimidated by the argument that a right action is impossible because it does not yield maximum profits, or that a wrong action is to be condoned because it pays. That philosophy is dead in human relations, said Leopold, and its funeral in land relations is long overdue. As Laurence Pringle points out, for many people the word conservation brings to mind dull textbook pictures of forest fires, or of ways to stop soil erosion. Not so long ago, these were some of the main concerns of conservationists. But as man's environment has changed, so has the meaning of conservation. The old problems remain, but new and more serious ones are with us. Today conservationists are concerned with the quality of life for all people, now and in the future. Just as preventing forest fires is a concern of conservation, so too is preventing noise in a city. Conservation problems are no longer "out there," seeming to affect only farmers and foresters. They affect everyone, and everything, living on earth.

We have to begin to ask questions, challenge ideas, and look at old problems in new ways. Should your community keep growing and growing, trying to attract new businesses and industries? Or should the growth be slowed or even stopped, allowing plenty of open space for the citizens? Should laws forbid the reuse of bottles, or should bottles and other containers be used over and over again? Should all the valuable minerals and other substances in sewage be sent downstream to the oceans or should they be returned to the land, to be recycled by nature? Sooner or later, hard questions like these must be answered. Some people still believe that man can escape from the earth if conditions here become unlivable. But there is no other place in our solar system where man can survive wtihout depending on supplies brought from the earth. Like it or not, we are tied to this planet.

There is a grand design. Wherever we look in nature, the same process appears to be taking place, and in the same direction. The elemental particles join to form atoms, which form molecules, which form compounds, which form communities of ever greater complexity. At every stage, some energy is "lost" or "traded" for a more complicated structure. The more complex the structure or system, the more possibilities it has for forming new combinations; its capacity for change grows. All the world, physical as well as biological, seems to be engaged in the same grand enterprise. Man is one, but only one, phase of the trek of all things through time. But as far as we know, only he is aware of this, and with awareness, assumes responsibility for the world in which he finds himself. It should be a joyous awareness, and a responsibility for which love is the only adequate adjective. This is the essence of the concept of the priest Teilhard de Chardin.

'Tis not too late to seek a better world. We are all dismayed at the scope and depth of our environmental problems—the four "Ps": pollution, population, pesticides, poverty. But time has not run out yet. Change is natural. The field that yesterday was a ragweed patch tomorrow will be a grove of aspen and after that a forest of maples. The geese that leave us in November return in March. So we can take it as a concept of ecology that improvement is possible, if we will but develop a will and a way to conserve as efficiently as we have consumed, to love as deeply as we fear.

Ecology is everybody's thing. Much modern science all too often seems to be a no-layman's land of abstruse concepts and unwieldy laboratory equipment, of arcane processes and complex formulas, of particles too small to be seen, and of power too great to be controlled. Not so, ecology. Ecology is where we live—in backyards, on boulevards, in parks and apartments, in fields and factories, in swamps and suburbs, in mountains and

metropolis. And everybody can practice ecology. It enriches our
pleasures and perceptions alike. It can stir us to respect and love.

"D" Stands for Dogma

Out of the ABCs of ecology has come a new set of tenets or
beliefs, held increasingly by the youth of the country. Grady
Clay, landscape architect, calls these new principles *Eco-logic*.
They can be summarized as follows:

1. Everything is connected to everything else. War in Indo-
china is a part of the Indiana ecosystem.

2. The earth is a delicate, closed life-support system that cannot tolerate unlimited growth and its wastes. Nothing can ex-
pand forever, and nothing ever really goes away.

3. All environments have a carrying capacity, a ceiling which,
like bank credit, cannot be exceeded without dire penalty.

4. Some environments are inviolable and must not be altered
by man.

5. The quality of life is more crucial than the quantity of
things produced. The question is *how* we survive, not mere survival.

6. For the environment to survive, society as a whole must
adopt constraints that may seem radical. We have met the enemy
and he is us.

In short, growing numbers of Americans now see their environment as a limited resource. They know that all environment is subject to man's influence, as well as the other way
around. They see all land, water, and air as invested with the
public interest. They believe the quality of the environment
should no longer be left to whoever happens to own a piece of
it. They are determined to have an increasing voice in decisions
about their environment. And that voice will become increasingly strident.

This is the ultimate in ecology: the translation of aesthetic
feelings and scientific knowledge into political action—*eco-tactics*.

2. Around the Ecology Calendar

SOME ECOLOGICAL NEWS DOESN'T MAKE THE PAPERS. THERE ARE times, at almost any season of the year, when we make news by creeping out of our close and crowded houses into the night or noon, shucking the accoutrements of custom, and partaking of what Emerson called "the medicinal enchantments" of nature.

These halcyon experiences may be looked for with a little more assurance in that pure autumn weather we call Indian Summer. There is one particular day in October when the countryside reaches its perfection, when the air and the land are in perfect tune. The day before, the oaks and maples were not quite at their umbre prime. The day after, a rain driving out of a dull sky will have stripped the leaves. But this particular day, immeasurably long, ineffably quiet, embraces broad hills and warm wide fields in a tempered light, casting an ancient spell on any who have chanced to be abroad in the magic moment.

But headline days can occur any time in outdoor America. The fall of snow in still winter air, preserving to each flake its perfect form, and embroidering in crystal the twigs and trails that ordinarily lead the anonymous life of the woods. A summer sunset sinking below the perch fleet to touch with gold the precipice of a mapled bluff. The odorous south wind of spring, waving with shadowy gust a field of soft young rye, and carrying with it the mysterious piping of migrating pintails.

Newspaper headlines typically are made in faraway places with strange-sounding names, but we can experience the inspiring, healing news of nature without leaving home. In every

41

landscape the punctuation point is the meeting of sky and land, and this is to be seen from Gopher Prairie as well as from the top of Mount McKinley. The stars at night glitter above the homeliest hillock in Raintree County with all the incandescence they shed on the Campagna. The clouds and colors of morning transfigure even back-yard boxelders.

Nature is seldom caught in undress; beauty can break out anywhere. The only real difference in landscapes is in the eye of the viewer. It is well, then, in our distracted days, to look for action where the action always is, in those odd moments of ordinary existence when reality and romance meet in outdoor vistas within our daily grasp.

Rightly done, a year of ecological observation begins in March, not in January. The outdoorsman doesn't pay much attention to man-made calendars. He goes by natural signs. It is in March in temperate America that the sap starts to rise in maple trees and human spirits.

March

Coming-Out Month

March is the month when the out-of-doors begins to come to life again, hesitantly yet surely. Flowers, birds, and humans shuck off their winter lethargy according to an ancient and immutable schedule. Before any wild blooms brave the still-frosty air of March, daffodils begin to spangle dun city lawns. To poets, the adventurous daffodil has been an inspiration of long standing.

Bryant called the daffodil "our doorside queen" as it pushed up "to spot with sunshine the early green." De Vere dubbed the daffodil the "love-star of the unbeloved March." Shakespeare wrote of "daffodils that come before the swallow dares, and take the winds of March with beauty." Perhaps the most famous tribute to the daffodil is that of Wordsworth, describing "a host of golden daffodils beside the lake, beneath the trees, fluttering and dancing in the breeze." But it doesn't take a poet to appreciate a March daffodil. We all respond with lifted hearts to this innocent yellow harbinger of spring.

The birds come back in March, too, waterfowl in the vanguard. Geese from Louisiana bayous bring new life to Dodge County cornfields and sloughs. Pintail and bluewinged teal stop over on their tour from South America to Canada. Whatever the geological or biological reasons for this annual spring migration, we know that the trigger is, if you'll pardon the expression, sex. Light stimulates development of the breeding glands of birds. As the season turns from winter to spring, the hours of daylight increase, and with it the sex drive. Finally, as breeding condition reaches full force, the birds strike out for nesting grounds in the north. The phenomenon can be produced by exposing captive birds to increased hours of artificial light. Fortunately, love life is triggered for different species at different weeks. Some migrants press northward in the footsteps of receding snows, while others sop up sunlight until late spring. The result is that the big migratory push is spread over several months, to the benefit of flyway traffic.

But the surest March sign of spring's impending return is furnished by man himself. Youngsters begin to rebel against trooping off to school in coats, preferring to shiver in sweaters. On the campus, student leaders begin to plan perennial spring uprisings like panty raids or sit-ins. Matrons are lured into shops to emerge wearing the unlikeliest of bonnets. The man of the house stomps around the back yard wielding a rake or an eight-iron.

True, in any March there is still to be experienced the last dregs of winter. A brief blizzard may blot out daffodils and defy geese for a day. The basketball season comes to a blazing climax in the state tournament. And you may have to keep your topcoat in the closet indefinitely. Yet the change is undeniably in the March atmosphere. As you drive home from work one evening, you are aware the sun is higher. When you get up one morning to fetch the paper off the front porch, you get a whiff of a languid south breeze. And then you know you have made it through another winter.

The Mysteries of March

A lot of people are discovering ecology these days, but its spirit isn't really new. In its simplest form it is a recognition of the relevance of nature, and this recognition has been a hallmark of English literature for many years. Indeed, expressions of kinship with the out-of-doors were a lot more common before 1870 than after. Interestingly, it was the mysterious month of March that prompted many ecological observations. Perhaps it is only natural that a month which marks the break from blizzard to balm would stimulate plowmen and poets alike.

One of the very earliest bits of English verse, so old the author remains unknown, observed that "March winds and April showers bringeth vo'th May flowers." Thomas Tusser, writing his "Five Hundreds Points of Good Husbandry" in the 1500s, said: "March dust to be sold, worth ransom of gold." The economic reasoning is lost on us of the twentieth century, but not the ecological. Mr. Chaucer was so taken with March as a turning-time that he called it "the month in which the world began, when God first maked man." A little later Hopkins cried, "Look! March-bloom!" The famous Bill Shakespeare was obviously out marking the change of seasons when he saw "daffodils that come before the swallow dares, and take the winds of March with beauty." Robert Browning adapted the spirit of March to one of his love sonnets: "How the March sun feels like May!"

For some reason or other, Alfred Lord Tennyson was particularly intrigued with the month of March. "All in the wild March morning I heard the angels call," he said, and went on to pen such ecological lines as these:

"More black than ashbuds in the front of March."

"When rosy plumelets tuft the larch, and rarely pipes the mounted thrush; or underneath the barren bush flits the sea-blue bird of March."

"Whenever a March-wind sighs, he sets the jewel-print of your feet in violets blue as your eyes."

By the age of industrialism, however, writers weren't so sure of their world. You can read quite a bit into the following passage from Lewis Carroll:

"'Have some wine,' the March Hare said in an encouraging

tone. Alice looked all 'round the table, but there was nothing on it but tea. 'I don't see any wine,' she remarked. 'There isn't any,' said the March Hare."

Science Catches Up With Spring

I suppose it was only a question of time before science overtook an old-fashioned symbol of spring. I guess it was bound to happen sooner or later. But some of us are still not sure we like the new streamlined way of making maple sugar.

Time was when maple sugaring was as much a spring ceremony in the life of a small boy as the state highschool basketball tournament or Easter. It all began in those first shy days of late March when the sap starts to rise in maple trees and human spirits. With a little auger you bored a hole in the tree trunk. The hole would begin to weep immediately and you would tentatively taste the sap as if it were communion wine. Into the hole you would push a twig of elderberry out of which you had poked the pitch to form a pipestem. Below the pipe you would hang a tin bucket to catch the dripping sap. Each morning and each evening you would make the rounds of your buckets, pouring the precious sap into a pail to be carried home to the kitchen.

They don't do it that way any more. They have a network of long plastic tubes which transports the sap from tree to tank untouched by human hands or spring spirit.

Once you had collected enough sap to fill a washtub, you came to the central point of the maple syrup ceremony. You put the tubfull of sap on the back of the kitchen range and you boiled and boiled and boiled. Gradually the sweet aroma filled the whole house. If you were making syrup you took the tub off the stove before the liquid crystalized. If you were making candy, you boiled some more. Knowing when to quit was an art.

They don't do it that way any more, either. A radical method of making maple syrup that eliminates most of the boiling operation has been developed by chemists as a spin-off from research aimed at recovery of potable drinking water from sea water. Called the "reverse osmosis" system, the new method involved pushing maple syrup through a semi-permeable membrane under high pressure. In other words, you get rid of the water in the sap by straining it out instead of by boiling it off in the form of steam. The new method is faster, cheaper, and surer, but the aroma has been taken out of the process, and with it a delightful symbol of spring.

Fortunately science has not yet caught up with all the signs of the vernal equinox. If you drive up to a river valley in March, you can watch the ice go out of the river with a primeval grinding and roaring. If you listen in the dead of night, you can hear the querulous honking of Canada geese as the great birds come back north from Baton Rouge. And if you look in many a city back yard, you can see a male of the human species as he takes his first rusty swings with a nine-iron.

Quiet: Ducks at Work

If mallards are in short supply these recent years, it's probably through no fault of their women-folk. Hounded by drought-stricken marshes in the north and predator-plagued shorelines farther south, the mallard hen remains one of the most persistent nesters of all waterfowl. Dr. George Burger, manager of the Mc-Graw Wildlife Area in Illinois, tells a classic story of mother obstinance in the face of odds.

In late March the heroine hen of this story set up house-keeping in a specially designed nestbox on a remote pond. By the end of April she was incubating a mammoth clutch of 20 eggs. One day in May, 12 downy fluffballs drifted to the pond from the nestbox and bobbed away after Momma. In 24 hours there were no more fluffballs—thanks to assorted crows, turtles, and so on.

Momma? Mad, maybe, but not dismayed. There were eight eggs left in the nest and by the next morning she was back to them with her life-giving warmth. By all odds they should have become infertile. However, patience and persistence were rewarded in a couple of days with four more fluffballs. But predators are persistent, too. The next morning the four were two. Then one. Then none.

This looked like the end to Biologist Burger. The hens of

most duck species would holler uncle. Even many, many mallards would have been content to retire to the reeds to molt and feed. But not our gal. It was back to the old nestbox, where she tidied things up a bit and laid eight fresh eggs. It was June when she began anew; July by the time she was incubating; July when heat threatened the tiny embryos.

It seemed unlikely that the eggs would hatch. Even if they did, wouldn't the ducklings disappear as they had twice before? Could the dogged duck make it against such odds? She could, and she did. All eight of those eggs hatched. Half of the ducklings vanished, but four youngsters made it, to migrate away with a mighty proud—and probably mighty pooped—Momma.

Many waterfowl species may be in trouble these days. But if the heroine of this story and her sisters have anything to say about it, the mallard, for one, is a long way from defeat.

Spring Sideroads to Squalor

One Saturday there was a faint hint of Spring in the early March air, so we left the city behind for the first country excursion of the season.

It is getting harder and harder to leave the city behind. To the west, on what only yesterday was prime cornland, there now sprawls the interminable skeleton of a mammoth shopping center to be, scarring the skyline with steel girders. To the south and east, where yesterday were woodlots and ravines interrupted only by an occasional farmhouse, now hives of three-bedroom ranches are beginning to dot the landscape, and new corner taverns testify to plenty if not to progress.

This may still be the land of the free and the home of bravado, but the beauty that was hers is waning.

If you keep going long enough and far enough away from the city you will eventually get to fairly open country and to country towns. But now those country towns are growing, too. One Saturday afternoon out at Lake Mills there wasn't any room in the parking lot at Eddie Dettman's spa, and the fishermen on Rock Lake marsh were thick enough to set the ice a-groaning. The pollution of population was particularly evident around our marsh shack. Some time during the winter snowmobiles had made a racetrack out of our terrace, and to make it easier to get their machines across the railroad track they had ripped up our picnic table and used the planks as girders. I would submit the bill to my insurance company, except that my vandalism clause has been cancelled because of the high risk. Where are the gentleman sportsmen of yesterday?

Oh, well, we thought, we can still find one refuge from a world that is too much with us. We will hike in to Bean's Lake. That was a mistake. Bean's Lake was once a gem of a pothole, hidden in the tamaracks, accessible only by foot. Now a road of sorts has been ripped through prairie groves and pitcher plants to accommodate—of all things—a town dump. Just beyond is a "Lots for Sale" sign that spells the imminent end of solitude.

There was one bright note to the day—the clamor of Canadas high in the cold March sky, the leaders calling the tired stragglers on, their wild chant echoing over bogs and billboards—a magnificent chorus of disdain for men who have squandered the spirit of the swamps.

April

Spring Is for Poets

The coming of April and spring have been to English and American romantic poets an experience of compelling inspiration. For twentieth-century writers, things are tougher.

"Oh, to be in England, now that April's there," wrote Robert Browning a hundred or so years ago. "Whoever wakes in England sees some morning, unaware that the lowest boughs and the brushwood sheaf round the elmtree bole are in tiny leaf." Today, the lowest boughs along my Iowa County sideroad aren't breaking into leaf, and they won't. They have been sprayed to death by highway crews who are paid to place more value on a clean right-of-way than on natural beauty.

"It is a beauteous evening, calm and free; the holy time is quiet as a nun, breathless with adoration," wrote William Wordsworth from his English lake country about 1800. Wordsworth obviously never had to contend with power boats. A spring sunset on Rock Lake today is about as quiet as the noon-hour on Capital Square. Viewing London from Westminster Bridge early one spring day, Wordsworth was able to say that "this city now doth like a garment wear the beauty of the morning . . . all bright and glittering in the smokeless air." You can't say that about Los Angeles now, or even about Milwaukee.

I once made the mistake of visiting the Wye River where, in the spring of 1798, Wordsworth wrote his classic "Lines Composed a Few Miles Above Tintern Abbey." To Wordsworth, the sylvan Wye with its soft inland murmur was the embodiment of a landscape from which he could always draw a sense of sublime purpose. To me, the Wye turned out to be as inspiring as the silt-stained Pecotonica.

It was even a worse mistake a year or so ago to visit Walden Pond, in Massachusetts, where on spring days in the 1840s Emerson and Thoreau saw "reflections of trees and flowers in glassy lakes" and were thus "led in triumph by nature" to express a new American ethic. Far from being any sort of retreat today, Walden Pond is within earshot of a Concord subdivision, and it now comes complete with a busy bathing beach and a souvenir stand, where for a hundred dollars you can buy an "authorized reproduction" of Thoreau's field notes.

Yet the voices of these April poets have not really been stilled. They spoke recently through a federal grant of $511,000 which the state will match to preserve 59 miles of Wisconsin's Wolf River, and in Senator Gaylord Nelson's act that another 29 miles of the Wolf be included in a new National Scenic Rivers System. Writing an April poem in 1990 will be tough, but it will be possible if we can keep on thus preserving those vistas where nature can continue to "impress with quietness and beauty" and "feed with lofty thoughts."

An Ecological Easter

It has become very fashionable in recent days in the circles of the new environmentalists to blame our current conservation problems on "the Judeo-Christian tradition." This is despite the fact that other cultures are not exactly models of a state of harmony between man and land. The definitive statement is an essay on "the historical roots of our ecological crisis" by a distinguished scholar in an issue of *Science* magazine, which has been reprinted with gusto in several heady new paperbacks.

There is a good deal to the theory, you have to admit. After all, on the first page of Genesis we are told to: "Be fruitful and multiply, and fill the earth and subdue it; and have dominion over the fish of the sea, and over the fowl of the air, and over all the earth, and over every creeping thing that creepeth upon the earth."

On the other hand, from hindsight we can see that this wasn't too bad a marching order at the time it was issued. It fit the needs of nomadic tribes looking for green pastures and still waters in an underdeveloped world. The trouble is the dictum doesn't fit modern times. But that's not the fault of the Bible. The fault lies with us. We haven't read the last chapters.

What is demanded of the human animal today? One quality is humility—humility in our relationships to our environment. And where do you find a more striking lesson in humility than in the story of a God who assumed flesh, lay helpless in a manger, and hung dying on scaffold surrounded by thieves? Another required quality is neighborliness—neighborliness towards all our fellow men and toward all those things that creepeth upon the earth. And where do you find a more telling exposition of neighborliness than in the parable of the good Samaritan? A third quality is hope. And where do you find a more eloquent testimonial to hope than in the Easter story of how the stone was rolled away from the tomb?

Our traditions tell of the Joshuas with their trumpets and tumbrels. They tell of the Jeremiahs with their voices of gloom and doom. But they also tell of the Johns who prophesied that "there shall be no night there." If the Judeo-Christian tradition is the root of our ecological crisis it can also be the font of our environmental recrudescence.

Sportsmen Should Sound Off

Every April, in each county of my state, sportsmen gather for their annual "fish and game" hearings. In few other states do hunters and fishermen have such a formalized chance to make themselves heard and felt, but the sound that emerges each spring in Wisconsin is seldom significant. It is a safe bet that at most county hearings the tedious details of hunting and fishing regulations will be argued line by line, and probably with only the customary tiny minority of all licensed sportsmen present. It is also a fair guess that most resource management issues that really count will be absent in the discussions. Some of the items on the agenda are sometimes not without substance: motor trolling, goose tags, deer hunting changes, and so on. But there's often nothing about crucial water quality standards, for example.

These yearly hearings are part of Wisconsin's statutory machinery for sounding out grassroots views on conservation. From the county meetings delegates are elected to a statewide meeting of the Wisconsin Conservation Congress. The Congress in turns makes recommendations to the Natural Resources Commission, which then sets the rules governing hunting and fishing. The hearings and the Congress, of course, *must* consider proposed changes in harvesting regulations. The system was set up in 1934 with this purpose. But there is nothing to prevent the Natural Resources Department from adding to the agenda a discussion of problems more basic than the length of a brook trout, and there is nothing to prevent sportsmen from sounding off on resource matters that really bother them, like the galloping trend of marsh drainage and woodlot destruction.

It may be instructive to take a look at the topics discussed at a recent congress of outdoorsmen in the state of Kentucky. The bag limit on bluegills wasn't even mentioned. Instead, there was an attempt to answer such fundamental questions as these:

How can we reduce the bad effects of some chemical pesticides on wildlife and wildlife habitat? How can we open up more lands to public hunting, particularly around population centers?

How can we encourage on soil-bank lands those practices that aid rather than harm wildlife production? How can we discourage the continued draining of wetlands?

Can we develop a land-use classification system that doesn't relegate wildlife to rocky outcrops and wastelands? How can we slow down or halt streambank cutting, stream channel straightening, and bottomland clearing?

How can we save pockets of wildlife habitat in the face of highway and subdivision bulldozers? How can we begin to make a dent on the sweeping problem of water pollution?

How do we protect our last stands of native flora and fauna from the onslaughts of picnic tables? How do we spread the costs of outdoor recreation facilities fairly among all the beneficiaries?

These are the sorts of questions agitating other people, too. Sportsmen have in the Congress system a good April platform. It should be used for something besides haggling over fish and game rationing.

Sometimes it is. Newspaper reporters covering a more recent meeting of the Wisconsin Conservation Congress had a tough assignment. Unlike many past stormy sessions, there was no conflict, no drama. The story was all between the lines. Sitting there listening to the resolutions being passed without a murmur, you could hardly believe that this was the Congress of yore.

Take the proposal for a two-party deer hunting permit, for example. This plan would give pairs of hunters in selected areas permission to take a doe as well as two bucks. For years the Wisconsin Conservation boys have been trying to persuade the hunters of the state that a bucks-only law is not sacrosanct and that the problem in many regions is one of too many deer rather than too few. For years the Congress has resisted the movement to enlightened deer management. Gradually, however, the climate has changed. In 1968 the last step was taken without one voice raised in dissent.

Or take the proposal to permit managed burning of certain areas in the north to encourage sharptailed grouse. For years the lighted match has been the symbol of all that is unholy. Now the Congress has reached a level of ecological sophistication wherein it unanimously approved deliberate fire-setting. Even more important, perhaps, the Conservation Congressmen spent very little time debating mickey-mouse changes in hunting and fishing regulations. Instead, they leveled their guns against DDT spraying, against dams on the Wolf River, for more access to public waters, and for more acquisition of outdoor recreation areas—significant topics rather than minor modifications. The Congressmen even turned down an experimental shooting season on mourning doves, which is something hunters aren't supposed to do.

What has happened, of course, is a very slow but very sure triumph for conservation information and education in particular and the democratic process in general. The theory of the Congress is simply that rational men, given the alternatives, will choose the sounder course. Many times in the past it has seemed that this theory just wasn't working. But professional and lay conservation leaders didn't give up. And they have finally been rewarded by a Congress capable at times of facing basic biological facts and acting with intelligence.

Spring Breaks Through Again

My *Encyclopaedia Britannica* consists of 24 fat volumes that supposedly are the detailed font of everything worth knowing. Hence it was something of a shock the other day to discover that all the *EB* has to say about Spring is that it is "the season of the year in which plant life begins to bud and shoot." Immediately preceding are two pages on "Spraying and Dusting Machinery," and immediately following are five pages on "Staff, Military." But only one sentence on Spring! Sometimes you have to wonder about scholarly perspective.

Poets have consistently displayed more sense. In the *Oxford Dictionary of Quotations* there are only a handful of references to Winter, but it takes over a page just to list the citations about Spring.

Perhaps one of the most famous references to Spring is Tennyson's line about how "in the Spring a young man's fancy lightly turns to thoughts of love." The preceding line is much less well-known but much more striking: "In the Spring a livelier iris changes on the burnished dove." It may surprise you to learn that somebody actually did write once about "the flowers that bloom in the Spring, tra la." He was Sir William Schwenck Gilbert, better known as the first half of Gilbert and Sullivan. Shakespeare took a good deal of inspiration from Spring, "when proud-pied April, dressed in all his trim, has put a spirit of youth in everything." The illustrious romantic poets were equally entranced. Shelley described the season and its winds—"the azure sister of the Spring." Coleridge said it was the time when "all Nature seems at work." And Keats referred to "the songs of Spring."

A whole host of lesser poets have played all the changes on Spring. Swinburne, for instance: "For winter's rains and ruins are over, and all the season of snows and sins." Thomson: "Come, gentle Spring; ethereal mildness, come!" And Thompson: "Spring is come here with his world-wandering breath, and all things are made young." Not all poets have been happy about Spring, of course. Eliot expressed the feeling that "April is the cruelest month, breeding lilacs out of the dead earth, mixing misery and desire." But Browning set the more popular tone when he said that "the year's at the Spring, and all's right with the world." And Allingham called the first week of Spring "a thing to remember for years."

On any April Sunday most of us would agree.

May

Golden Memories

About a May morning there is a special quality that can turn time backward. Gentle air, filtered sunlight, scented grasses, and bird song inspire mystic memories of bygone springs when the world was new. Perchance, those memories involve trout fishing. Samuel Johnson may have said that a fishing rod is "a stick with a hook at one end and a fool at the other." But many of us from our youths have preferred Izaak Walton's dictum that "angling is an art worthy of the knowledge and practice of a wise man."

Trout fishing in the '20s meant pursuing quarry with the romantic titles of Lock Leven and Von Behr. Through time and interbreeding, these varieties have now lost their immigrant Scotch and German traits, and the fish is presently known by the pedestrian name of brown trout. There is nothing common, however, about the sporting qualities of the brown, as fishermen discover as they open the season each May. It could well have been a brown trout to which Shakespeare referred when he said that "the pleasantest angling is to see the fish cut with her golden oars the silver stream." The brown is not really brown. He is golden, distinguished by orange or brown spots on the side and by an orange tinge of color in the small, fleshy adipose fin just behind the main dorsal.

Before we knew much about trout management, we used to plant browns as tiny fry in the fond hope that they would survive. Very few did. Then we switched to stocking legal-length browns each spring just before the season opened. They had a rearing-pond pallor and were pretty vulnerable to lures. Now we know better, thanks to a good deal of fisheries research. We plant fingerling browns in the summer and fall. They grow to sporting size and habits on natural fodder, and provide good angling opportunities for several seasons.

Purist trout fishermen operate on streams with artificial flies bearing such exotic names as Royal Coachman and Parmachene Belle. But the biggest strings of trout will be taken at night by small boys of all ages soaking nightcrawlers—in Thomson's words, "beneath the tangled roots of pendent trees." For getting back behind the years there is nothing like a May evening, a bamboo pole, and a squiggly worm. The stream bank may bear the marks of a Public Law 566 watershed improvement project; the water may have gone through two or three Cross Plains kitchen sinks; and the trout himself may bear a tag reading, "Manufactured at Nevin State Fish Hatchery." But the tug on your line is a message straight from a pleasant past.

Now is the Time for Long-Looking

Now is that magic time of the year when man and nature can be most in tune. The inhospitable winds of winter are but

memories. Yet to come is the almost obscene flourish of summer
vegetation. In May our bodies and minds can strike a balance
with the out-of-doors.

But to do so we must have open space—that priceless vista
where trees and plains meet an untrammeled sky. This need for
a view, for the long look, may be a peculiarly American require-
ment. Europeans do not seem to have cultivated it. They have
gone in more for temples and formal gardens. But not Americans. The craving for the long look brought us here in the first
place, and sent thousands of Lewises and Clarks beyond one
range after another, until finally we stood, like Balboa, silent
upon a peak in Darien. My Greatgrandfather Jones came from
New Jersey to Wisconsin in the early 1800s, looking for the long
look. He stayed only briefly, and then went searching still across
the wide Missouri and beyond, like so many others of his era.
He and his kind won the west while they were looking, and
set aside some parks and preserves so we could look, too.

Recently George Gallup discovered just how precious open
space is. Polling Americans about conservation, he found that
although the bulk of his interviewees lived in cities, they didn't
really like to; nine out of ten said more land should be set
aside; and most of them said they would pay extra taxes for it.

To really look beyond the ranges, of course, you have to get
on a high peak on the Kenai, where your eyes can encompass a
whole world that still knows men not. But you can get a little
taste of long-looking on a May evening even in the city by
plunging off the beaten tracks to a place like a park and soaking
yourself like a muskrat in the spell of woodland spring. Old,
open-grown bur oaks will be weaving their ancient lacework
against the horizon, at your feet will be the fragile flora of a
remnant savanna, and hidden lagoons will be loud with frog
voices.

We stood the other night, listening for the "peenting" of woodcock, as a pair of teal threaded their way through the dusk and
the willows to land with so soft a splash in waters that have
known their kind since Pierre Radisson first came long-looking
to the valley of the Mississippi. It was only for an instant. Inevitably we had to consign ourselves again to the cacophony of

a bulging beltline highway. But for a magic May moment we took the long look and preserved it on an island of the mind.

The Plaint of the Red Plaid Shirt

Bird watchers and bass fishermen, when you choose your spring wardrobe, beware! A red plaid shirt is no longer the exclusive mark of the outdoor fan. It has been taken over by the hippies. Since I don't know when, the red plaid shirt has been the unofficial uniform of the outdoorsman, a secret sign that the wearer is a member of the fraternity of nature. Now the red plaid shirt is a sign of urban revolt.

Whether they actually wore them, I don't know, but in paintings the French voyageurs are always pictured sporting red plaid shirts as they plied the shores of Lake Superior. Paul Bunyon, that mythical lumberman whose St. Croix exploits are only a little less sensational than the performance of Bart Starr, is always shown in a red plaid shirt. Even before the law said he had to wear red, the Wisconsin deer hunter garlanded himself in a red plaid shirt. The same garment is the traditional mark of the musky man.

A generation ago, when some of us were in high school during the depression, the red plaid shirt did double duty. It was not only as inexpensive a warm getup as you could buy (as I remember it, you could purchase a good red plaid cotton flannel in those days for 79 cents), it marked its wearers as the rugged outdoor type who would really rather be in a duck blind than in class. The principal made noises about our changing into something more appropriate, but we didn't pay any attention, and he didn't mean it, anyway. Just before the war, on the University campus the red plaid shirt was the required uniform for all members of the Hoofer's Hiking Club, and for anybody attending the annual Winter Carnival dance. Today, if you pay a visit to a small-town spa, the red plaid shirt is still to be seen as the mark of the local hunter and fisherman. Or, if you should chance to join the Madison Audubon Society on one of its Sunday morning bird walks, you had better wear a red plaid shirt if you don't

wish to feel conspicuous. The same thing goes if you attend a
conservation education conference at Trees for Tomorrow or a
Nature Conservancy hike in the Arboretum. Your fellow ecology
buffs will all be attired in red plaid shirts, befitting their status
as knights of the John Muir roundtable.

But the days of the red plaid shirt as the mark of the outdoor
fan may be numbered. The red plaid shirt has been taken over
by the New Left on our college campuses. Attend a sit-in on
Bascom Hill, and the participants will be wearing red plaid
shirts. Watch a picket line on the Library Mall, and the Mario
Savios of Madison will be wearing red plaid shirts. Go to a rally
in the Rathskeller, and the leaders will be known by their red
plaid shirts.

It may be poetic justice that we outdoor fans have been thus
robbed of our distinctive attire. After all, our "back to nature"
movement has always represented something of an escapist cult,
albeit a part-time diversion. Now our hippie friends have made
escapism of a different sort a full-time pursuit. They probably
view the red plaid shirt in the same light we do—as the antithesis
of the grey flannel suit.

June

June Is a Jumping Bass

Some dates stick in your mind like a cocklebur to a hunting
shirt. One such date is June 20th, which used to mark the annual
opening of the bass season. That was back when we thought we
had to protect brood bass on their spawning beds if we were to
save the sport. Now we think we know better. We recognize it's
just as easy to have too many little bass as too few, so we open
the season in April. For some of us, however, mid-June shall be
forever associated with the proper time to switch from trout and
pike to bass.

As all wielders of rod and line know, the bass is a very special
kind of fighter. A trout is a fancy-dan. He will feint and spar,
back-peddle around the ring, and then close in with a flurry of
blows to the head. A northern pike, on the other hand, is a street

brawler. He has no finesse, only brute strength. He likes to swing hay-makers from left field, or punch you in the kidneys in the clinch. The bass is something else again, a finny combination of Gene Tunney and Jack Dempsey. He can engage in the fanciest footwork to throw you off guard, and then deliver a massive blow to your solar-plexus when you aren't looking. He will sulk among the stumps one moment, and come churning up in a wild leap the next.

If you are like me, you cut your eye-teeth on small-stream smallmouths. You tied on your first artificial lure—a Skinner spoon —and skittered it beneath shelving banks and over rocky runs, using a cane pole three times your height. And you have never been the same since a bass as big and black as a medieval knight smashed into your primitive tackle. Now you have probably graduated to lake largemouths, which you pursue with more sophisticated gear, twitching marvelously lifelike plastic frogs among the coontail, or sending plug minnows probing down where sandbars drop off into cool green depths. The result is the same: a bass striking like the burst of a hand grenade to leave you as committed as ever to the pursuit of "the gamest fish that swims."

Whatever the law books say, June is still the time you turn to bass fishing. Maybe it is because of the setting: along a conversing creek, where the scents of mint and clover spread in the first summer heat to provide an angler's aphrodisiac; or on a quiet bay, where terns dip down to dimple the waters, and pond lilies stay abloom far into the evening. It is in such peaceful surroundings that the black bass lurks, ready for his June jump into your heart.

P-R Has a Birthday

To some folks, PR may stand for "public relations," but to conservationists PR stands for "Pittman-Robertson," two congressmen who have lent their names to a wildlife restoration program that celebrates its birthday each June. Under the terms of the historic PR act passed in 1937, the proceeds from a special 11 percent tax on sporting arms and ammunition go into a special fund for distribution to the states. Each state adds 25 percent more and uses the combined money for wildlife habitat acquisition and wildlife research. It is not too much to say that PR funds have made the difference in state wildlife conservation programs in the past 30 years.

Typical results of PR work are these:

Question: How effective is pheasant stocking? Research finding: depends on location; only in heavily-hunted areas or in areas with low wild pheasant populations do stocked cocks significantly raise kill by hunters; stocked hens produce few broods, so contribute very little to the bag. Conservation benefit: Sportsmen's clubs encouraged to stock cocks only in heavily hunted areas; sexed-chick program developed by state game farm; hen stocking cut back; farm operating costs reduced; hunters still bagging as many stocked cocks as before.

Question: How can deer hunting success be improved? Research finding: by gearing hunting seasons more closely to deer populations; better status information now continuously available through improved methods for inventorying deer numbers; projecting population trends, evaluating hunting seasons. Conservation benefit: development of unit management system, resulting in more deer taken, less range damage; in time will mean more stable deer harvest and population size.

But progress in wildlife conservation has depended on more than PR dollars; it has depended also on a dedicated corps of able game managers and researchers. Yet the real guest of honor at any PR birthday party should be the average hunter—the guy who has picked up the tab by paying that 11 percent tax on guns and bullets. All told, he has quietly contributed over $300 million to the federal PR treasury in 30 years. Meanwhile, under a similar Dingall-Johnson program, he has been paying a special tax on his fishing tackle. There is a lesson here for other consumers of outdoor resources. Maybe it's about time that the buyers of tents, trailers, fieldglasses, powerboats, chain saws, cottages, skis, scuba gear, and other accouterments of outdoor recreation start paying their share of the open space acquisition and development bill.

July

Fourth of July Fresco

July Fourth will find most folks celebrating Independence Day out-of-doors. It is fitting that this should be so. From the moment we dropped anchor in the shelter of Plymouth Rock, we have been children of wide open spaces. There was at hand the forest in all its primeval arrogance; beyond, great prairies, and mountains running to unknown oceans; a geography written on by no one, a history unmade, an inland empire to shape and fill. Trappers penetrated the wilderness, lead miners followed them to make a frontier, lumberjacks cleared the forests, farmers poured in after them, railways hacked their way across the savanna, brewers mixed hops and spring water to make Milwaukee famous. This is our heritage.

Little wonder that modern Americans possess an almost mystic yearning to get into the out-of-doors, and that they do so with particular fervor on July Fourth. Be we a Thoreau or a lathe operator, when we look for meaning in life we seek it not in ancient ruins or in canyons of a city but in a forest, by a lake, or at the edge of a river. The out-of-doors is our inspiration, an opportunity to recapture, if only on a picnic in a park, a sense of that magic potion of wide skies and free minds that is America.

You will find it good to get out into your state this month. You read and hear so much nowadays about how things have gone sour, and you start believing it, until you get out and see for yourself that they haven't, and that it is pretty much the same friendly, bold, naive, surging land it has always been. It is still an open land, full of space—the Badlands for instance—so much earth and rock and forest and water and sky that it makes you wonder a little about all the talk about a population explosion.

It is still a country of striking beauty: standing on a bank of the Mississippi and watching that green giant snake its way to Louisiana; riding to the top of the vista needle at the Dells and seeing the monumental carvings of the glaciers dwarf into utter insignificance the throngs of tourists; coming upon a hidden lake in the bucolic hills of the Kettle Moraine; experiencing the Main

Street of a small town in all its picture-postcard charm; or driving into the shoulder country of northeastern Minnesota's eroded mountains, each rise opening new vistas, until you can stop the car and get out and just stand in awe of the might of Lake Superior glistening in the sunset.

Perhaps what you will be struck with most in July, however, along with the beauty and the energy of the countryside, is a very simple fact: the normalness of the people you meet. We get so wrapped up in the country's problems that we tend to overlook normalcy nowadays. We accentuate the kooks. Actually we American people are pretty much what we always have been. We wake up each day, sniff the morning, and wonder what's in store; we dress, eat, work, think, argue, read, make love, hunt and fish, camp and hike, drive our boats and cars, worry about the bills, take pride in our children, try to outdo our fathers, cut our lawns, mend our houses—and finally sleep again with hope in our hearts for tomorrow.

That's what a Fourth-of-July outing is all about—rediscovering a land of beauty and excitement, of raw possibilities, of hard work and ingenuity, of dreams, of opportunity, of patriotism and pride.

Oh, Say Can You See?

When I was a boy, July 4th was celebrated outdoor-style in a rather simple way. At the crack of dawn we would shoot off firecrackers in our back yard. Then we would head for a prairie grove at Waldwick or Latto for a potluck community picnic, featuring a stirring address by Fighting Bob LaFollette and an equally long invocation by my father. After which we would sneak off to the nearest sucker stream with a cane pole and a bobber.

All that is considerably changed. Outdoor Recreation Model 1971 comes in a bewildering array of shapes and colors, many of them manufactured in Japan or at least in Detroit. At one end of the spectrum is the big boatrace, as staged recently in Madison, Wisconsin. It is a little hard for some of us to understand how 50,000 people can get their kicks out of watching a dozen hydroplanes make mincemeat of Monona, but suffice it to say they do. At the other end of the spectrum the same weekend the small cadre of folks who followed Biologist Jim Zimmerman around the Arboretum's restored prairie, drinking in the wonders of lady-slipper, Indian paintbrush, and spiderwort. The boat buffs probably don't understand the wildflower fanciers.

In between are the mushrooming numbers of recreationists who hie to such places as Governor Dodge State Park, where the Department of Natural Resources is turning what used to be a secluded hollow into a center of mass diversion; or to arty spots along Highway 23 between Dodgeville and Spring Green, where music and theater fans enjoy all sorts of performances, wander through curio shops, or try to sit in Frank Lloyd Wright-designed chairs. Still other folk just go fishin' on the Fourth, although the equipment they take along these days would be completely beyond the capabilities of a Model T.

All of these forms of modern Fourth-of-July recreation have one thing in common: they are the fruits of a technological age. No longer can we expect to serve the wants of our fellow citizens with baked beans on a buckboard under a bur oak. We have to plan, and build, and develop, and charge fees. Nature, in short, needs help. The many forms of modern Fourth-of-July recreation have another thing in common: they are essentially competitive. The powerboat jockey on Lake Wingra can effectively ruin the day for a Sunday afternoon hiker on its shores. And the ladyslipper preservationist can preempt woods that might otherwise harbor campers.

All of which suggests that the spirit of '76 will have to be reactivated. We have to be very concerned about the rights of many minorities. The consumers of outdoor recreation are not a single majority "public." They constitute congeries of minority "publics." To protect in some way the rights of each of these publics, we can't apply the classic slogan, "greatest good for the greatest number." That leads us down the path of a grey concensus satisfying nobody. We have to learn how to determine and meet many divergent demands and fit them into a pattern of zoning that will develop outdoor resources without needless destruction and protect them without penalizing proper use.

The Last Stand of Heroes

Everybody thought he was dead. He certainly had not been seen for a long time. Not since he was reported walking along the Mississippi River a few miles south of Hannibal, Missouri; or swinging a big 40-ounce bat in Yankee Stadium; or taking on a platoon of Boche single-handed in Belleau Woods. Everybody knew what he looked like because his picture was in most everybody's cedar chest. If it wasn't there, it was certainly bound in the archives of the Library of Congress, or simply engraved on the minds of millions.

He was the Great American Hero: Old Hickory. Buffalo Bill. Huckleberry Finn. The Sultan of Swat. The Railsplitter. The Son of the Wild Jackass. Sergeant York. He was that picaresque lone operator, larger than life, who set himself against the tide and rode into the sunset, trailing clouds of glory.

Walt Whitman once pronounced his eulogy: "As now taught, accepted, and carried out, are not the processes of culture rapidly creating . . . a man (who) loses himself in countless masses of adjustments, to be so shaped with reference to this, that, and the other, that the simply good and healthy and brave parts of him are reduced and clipped away?"

There is, indeed, little in our era hospitable to the Great American Lone Ranger. Most of our public figures come prepackaged in Brooks Brothers suits. They are adept at the politics of concensus. They muster about as much charisma as a kewpie doll. Even the stardom of the astronauts is vitiated by the teamwork behind their feats. More than anything else, the idea of "the team" seems to have made the American Hero obsolete. There are scientific research teams, offensive and defensive teams, fund-raising teams, diplomatic teams, sales teams, team teachers. For a time, sportsmen resisted the team approach, posing as reincarnations of Daniel Boone. Then came the pernicious doctrine of togetherness. When that modern Don Quixote, Bobby Kennedy, braved the gorges of the Colorado, Ethel and ten little Kennedys went along.

But the Great American Loner isn't really dead. You can still find him on a golf course in July. This may account for the rather sudden and sensational rise of the Arnold Palmers as folk

heroes. Arnold is not just another number on a well-oiled football machine. He is out there all alone, with nothing between him and disaster but his trusty putter. He is not an amorphous, anonymous scientist in an off-limits laboratory. He is a grimacing, glowing human being in the full glare of the TV cameras. The course he is attacking is in every sense a replica of the American frontier. There are forests, swamps, and deserts, great prairies, lagoons, and quiet meadows. He drives off from a tee as if he were embarking from Fort Dodge, and he strides down the fairway with no sign of fear for the savage traps lurking along the overland trail to California gold. In his wake is Arnie's Army, the hundreds of thousands of acolytes who each summer weekend pack their shooting irons, saddle their carts, and fare forth to do battle with 6,000 yards of wilderness. It is a pretty antiseptic wilderness. But it's all we've got left. It is the last stamping ground of the American hero.

August

The Sun-Tanned Face of Summer

There is a special spirit in the outdoor air of early August. It is almost as if Nature were aware of man's habits. Like hundreds of thousands of humans, she tends to take a vacation this time of year. You can see and feel her pause perceptibly. Past is July's green avalanche of maturing vegetation. Yet to come is September's war dance. Now is a quiet interlude.

August dawns are moist, and fog fingers hover in the river valleys and the coulees. Lawns and pastures are spangled with cobwebs. By noon, however, the sun will have burned through and the landscape lies hot and seared. Wherever you go, a summer drought will be leaving its scorching mark on upland pastures. Lowlands that in spring stood four-fifths under water will seem half-deserts now.

The temperature will be torrid by noon, and great thunderheads will mount in the afternoon sky. Only an occasional vagrant wind will wave the woods and sweep with shadowy gust a field of corn, inhabited by a legendary Indian Goddess, her limbs swathed in the nodding leaves. Goldenrod is yellowing brightly along the roadsides. Sweetcorn is plentiful. (Incidentally, a cob of corn ought to be pulled, husked, and put in the pot, like a brook trout, while it is still flapping.)

Bass, pike, and muskies are in retirement, lolling in lake and river depths, coming out to feed only at dawn or dusk. Trout lie wary and unresponsive in the gin-clear water of their streams. The strident voice of a young crow serves only to accentuate the silence of inactivity. Big bumblebees drone by. Around a sudden bend in a country road you come upon a flock of partridge pullets dusting themselves in the sparse gravel. They take off in a flight not yet purposeful.

A lowing herd winds slowly up the lane. A distant canning-factory whistle summons the evening shift to work, the lights begin to twinkle from the barns, the long day wanes, the pale moon climbs, crickets call 'round with many voices, and the insects inherit the earth. All night their humming chorus prevails, of voices so threadlike that only their mighty numbers make them audible. Flies carve incredible geometric arcs through the heavy air. Moths dance giddily in the light; mosquitoes sing high, thin malice; baffled bugs bump clumsily against cabin screens.

Yes, in August you are almost tempted to think that Nature has struck a balance, that some power has indeed commanded the seasonal moon to stand still. But beneath the lassitude and brooding the powerful forces of the turning year are immutably at work. By the end of the month there will come a change in the air, faint yet sure, a cool breeze riding down the celestial track from Canada. And you will rummage in the closet for a sweater.

Hidden Hordes

This is the time of the late-summer doldrums, of the dog days. The midday sun sears the land in a brassy haze. The woods, fields, and marshes are silent. Leaves wilt and droop in the listless air. At no other time, save in the sub-zero days of northern winter, does the outdoor world seem so still. The appearance is deceiving. Hidden by the lush summer vegetation, the land throbs with life. Wildlife populations are at their annual peak, ranks swollen by young ones hatched or born in earlier months.

There's little hint of this abundance at mid-day. But venture afield at dawn and you enter a different world. The air pulses with bird song. Roadsides and woods edges are alive with cotton-tails, in a stepladder of sizes; with tidy, bright-eyed families of quail; with cautious does and stilt-legged fawns. Here and there a hen pheasant or grouse doggedly incubates a late clutch of eggs, her earlier efforts thwarted by weather or nest-robbing varmints. If she succeeds, hunters in the fall will call her late-hatched brood "squealers," and some will mistakenly give her credit for raising two broods. The waterfowl marshes are strangely quiet, even in early morning. Gone is the riot of sound and color of the spring courtship displays. The brilliant plumage of the drakes has disappeared, replaced—in a molt unique among all birds—with drab, henlike feathers. Most of the drakes themselves have vanished from the marshes, banding together on more open waters before this molt deprives them of all flight feathers.

While the drakes seek the safety of open water, the hens, also molting and flightless, remain close to cover with their broods. For some, the peace will be disturbed by crews of waterfowl biologists, taking advantage of the molting period to round up, count, and band the earth-bound birds.

By mid-morning across the landscape the flood of life ebbs into the shade, for a long siesta through the heat of the day, to venture forth again with the lengthening shadows of late after-noon. Stuffing themselves on the abundance of late summer, the wild youngsters—all unconsciously—are racing against time. All too soon the days will be much shorter; the hint of frost will be in the air. Young wings must be strong, and new feathers full, to carry those who will migrate beyond the reach of winter. The youngsters who remain, the earth-bound mammals and the non-migrant birds, must develop strong bodies, well fortified with reserves of fat. Men throng the beaches, huddle in air-condi-tioned rooms, and cluster around fans. But in the dog days of August, the hidden hordes make ready for the harvest dance of fall.

September

I'm a Fall Guy

For everything, says the Book, there is a season. For many of us the season is fall. On the face of it, there may appear to be something pathological in our love for the season of dying. On the other hand, autumn comes and goes with such a flourish that we can scarcely be faulted for our devotion to a blaze of glory. For us fall guys, autumn is our Camelot, our shining hour.

There is, first of all, the sheer avalanche of color. The crimson of the maples, the evanescent gold of the hickories, the frosty purple of the asters, the candle yellow of the poplars, the scarlet of the sumac, the lavender of the oaks—all set against stark white birch trunks and soft green pine boughs. You can hardly encompass the riot of nature's brawl.

Then there are the activities of fall: the upland grouse hunt, with a russet bird rocketing away through the rasping trees; the marshland pheasant hunt, where an immigrant Chinaman matches with his own burnished breast and silvered wings the colors of his adopted home; the pothole duck hunt, where a long line of mallards sends down from great heights that susurrant sound of wings and chatter of conversation that brings goose pimples to the neck of the waterfowler; the Arboretum hike, where the haze from burning leaves across the lake infilters a prairie grove to become ghost smoke from Indian campfires; the living room evenings, with good red apples and nuts; football—in Camp Randall, where the crowd roars as 11 men in cardinal trot out onto what is now perpetually green turf, or on TV, where the incomparable Green Bay Packers sponsor séances in myriads of homes; or that last golf match, when your ball rolls an extra 30 yards on the baked fairways of Indian summer, and the greens lie strangely quiet in the still air.

You don't even have to engage in vigorous diversions to get a lift from autumn. A ride through the purpled hills will do it, or simply sticking your nose out the front door on that first frosty morning. Trout fishermen and gardeners may vote for spring, swimmers for summer, and skiers for winter. But as for me, give me fall, in the north, where the year ages with abandon, and I know that life can reach its zenith after 40.

Duck Hunter's Diary

The federal and state law books may say the duck season officially opens in October, but for all true waterfowlers the season really opens in September. Bob Ellerson and I usually perform the initial annual rites of the duck hunter, appropriately enough, on Labor Day weekend. The rite is known as "fixing up the blind."

No matter how sturdily you build a blind in the first place, the amount of annual maintenance is staggering. For example, the insidious power of ice in the grip of wind and wave is something to behold. It can be counted on to snap off a six-inch-thick tamarack post like it were a match stick. Changing water levels can render the floor of the blind either a foot under water or three feet above. A muskrat will likely have strewn his debris over the seat, or worse, undermined the whole structure. Coons and birds of prey will have left their refuse. Human predators will have defaced your signs. And last season's flourishing garlands of cattails and sedge will, of course, have been stripped clean by marsh gales.

After shoring up the basic edifice with due ceremony and swearing, you turn to the ritual of cutting, bundling, and tying on a new covering of camouflage. We have made a science of it. In the first place, you cut your vegetation at a spot well removed from the blind, so as not to disturb the natural surroundings. To facilitate forming sheaves, we construct a special saw-horse in which to cradle the bunches of grass and bullrushes while we secure them with binder twine. Each year we leave this frame cached in the marsh, and each year we have to build a new one. But we don't really care. It is all a part of the ceremony. So is tying the rushes to the snow-fence sides of the blind, an art known only to duck hunters. No woman arranging living-room draperies exercises such sophisticated care. From the perspective of every possible on-coming bird, the outline of the blind and its inhabitants must be perfectly concealed. To double-check our art-work, we row out and take a look before we tie on the last batch of bundles.

With his blind reconstructed and camouflaged, the duck hunter now engages in the most solemn part of the pre-season ritual. He assumes a shooting position in the blind, and he prays. The power of this pre-season prayer is equal only to the power of shoving ice. In no time at all, the sky is filled with ducks, imaginary yet no less real. Far off to the left we can make out the first flock as it rises off the lake and comes winnowing over the railroad tracks. They are mallards. Gracefully they trail along the far shore, swing past the south corner of the marsh, and then turn toward us as they see our mythical decoys. For a moment they hesitate, and then they are on us with a rush. We stand, aim our imaginary shotguns, and fire. Then we paddle out as if to pick up the downed birds, turn to admire our handiwork, and then slowly row away, meanwhile keeping an eye peeled for more imaginary mallards. The ceremony is complete. It is almost as compelling as opening day itself.

October

Autumn Crescendo

If things run true to form, a week in October will mark out-door America's finest hour. Other climes may see summer slip into winter with only a whimpering rain to mark the passage. But not in temperate North America. Here fall reigns with a flourish.

At first there was only a hint of autumn's annual orgasm: a rasping in the poplars, a subtle fragrance in the dawn as of poppy fumes and apple oozings, a splash of goldenrod along roadsides. But the hint was enough to set astir the heartbeats of fall. By the peak of Indian summer the cadence will have reached a crescendo: hills aflame with the scarlet and saffron of sumac and maple, marshes glowing with the frosty purple of asters and gentians, and overhead in the night sky the haunting call of migrating Canadas.

If you are fortunate you will be abroad on that one day in autumn when the outdoor world reaches a particular zenith, when foliage and temperature mount to a pinnacle of warmth before beginning their long skid down to December. There is no wind, the only sound the strident voice of a seven-year cicada, or the measured "plop" of a falling hickory nut. A delicate haze hangs over the yellow fields and golden woods. An even more delicate perfume is diffused in the soft air. The still sky is an untroubled blue. The ground is elastic under your feet. You breathe tranquilly, but there is a strange tremor in your soul. Your mind can scarcely encompass the ineffable beauty and peace of the hour. You laugh and you cry.

Sooner or later after this perfect day there will come a turn. Perhaps there is a three-day driving rain that strips the leaves from brown branches. Perhaps it is a killing frost that blights reeds and grasses. On a late autumn evening, as you stop a moment to savor the weather, there is a cold, raw edge to the rising wind. You catch a harshness in the swish of tree limbs by the house corner.

Fall has come and gone. But while she is here, as Browning put it, the American earth is "crammed with heaven."

That Half Hour Before Sunrise

October duck shooting starts a half hour before sunrise each morning instead of at sunrise. Anybody who has never been in a duck blind can possibly appreciate the importance of that crucial half hour. The fact that the hunting is better in those 30 minutes really has very little to do with it. The whole thing is a matter of principle. Duck hunting, you see, is at heart not so much a physical exercise as it is an emotional experience, in which the stage setting is at least as important as the action. There is a script that must be followed, else the play loses its point:

A northwest wind swishing around the corners of your marsh shack and whisking down the chimney to send little puffs of woodsmoke eddying around the room . . . the faint call of migrating Canadas penetrating the gusty night to cause electric

shocks to run up and down your spine . . . lumpy sacks of decoys
with tangled anchor cords . . . baggy shell vests . . . frost-seared
hay on the floor of olive-drab boats . . . clumsy hip boots . . . the
ear-shattering snap of a shotgun shell being rammed home into
a magazine in the silence of the swamp—all mixed mysteriously
together under a pre-dawn sky.

It simply must be a pre-dawn sky. There is no thrill in load-
ing your skiff if you can see what you're doing. The science
comes in stowing your gear in the terrible dark that precedes
the light. There is no challenge in finding your blind unless you
guide on the wind and the stars. There is no kick in tossing out
your blocks after daybreak. The knack lies in placing them by
feel and smell. There is no lift to hunching down behind the
cattails if the sky holds no secrets. It is a striking phenomenon,
the duck hunter's sky. Darkness grapples with dawn on a mat
of low clouds. You peer into the void, and there is nothing but
night. Then suddenly the light comes, and far, far off is a line
of moving waterfowl. No other moment renders civilized man so
much a part of his wild heritage.

Someday, in the marshes of the Milky Way, duck hunters who
make the grade will write their own perpetual hunting regulations. In keeping with the policies of Heaven, the only legal
birds may well be helldivers. But you can be sure the shooting
will always start a half hour before sunrise.

November

He Called It Conservation

In November 1898 a tall, gangling young man of 33 accepted
a post as head of the U.S. Forestry Office, a tiny, dead-on-its-
feet agency in Washington. He was Gifford Pinchot, forester.
Twelve hectic years later, when Pinchot left Washington, the
U.S. Forest Service had been formed, a total of 194 million acres
of public land had been set aside as national forests, sound forestry practices were being spread, and the term "conservation"
had been born more than six decades ago, in 1907.

Pinchot was bubbling with ideas for the better management

of public lands. He was disgusted by the habits of the great financial barons, who as he said, "stole, plundered, and captured public resources like they were parts of a chess game." He personified the populist revolt against monopolistic resource waste, and he found a monumental ally in President Teddy Roosevelt. The two men gradually began to realize there should be a single core of husbandry to the work of all the emerging agencies and bureaus involved in the management of natural resources. But the concept was nameless. What should they call it?

"I knew that large organized areas of government forest lands in India were named conservancies and the foresters in charge of them conservators," Pinchot later wrote. "I proposed that we apply a new meaning to a word in the dictionary and christen our new policy 'conservation.' I put that term up to T.R. and he approved instantly. So the child was named."

Pinchot was an anathema to entrepreneurs and state chambers of commerce who thought conservation ought to mean the "development" of public lands, waters, and minerals for private profit. They finally forced him out of office in 1910. Pinchot was also an anathema to the naturalists who thought conservation meant the preservation of natural wonders from any intrusion. Pinchot opposed John Muir, for example, on the issue of a dam that would flood part of the Yosemite; and he made sure Muir wasn't even invited to the celebrated White House Conference on Conservation in 1908.

There continued to be these two competing doctrines of conservation abroad in the land. The first, what has been called "the gospel of efficiency," has never been better expressed than by Pinchot himself in his dictum that land and water use should be governed by a concern for "the greatest benefit for the greatest number of people for the longest possible time." This principle runs hard up against a second view of conservation, what might be called the Muir-Leopold theses, that what we should really seek is "a state of harmony between man and land," in the attainment of which we must recognize that land and water, as well as people, have certain unalienable rights.

The term Gifford Pinchot invented 60 years ago has taken a place along with "flag," "home," and "mother" in the litany of American catchwords, perhaps because, as William Howard Taft once said, "Everybody's for conservation, even if they don't agree what it is."

Thanks, Man

In this business of conservation nowadays you are sure to get yourself read or listened to if you pick up the familiar cudgel and beat the daylights out of Man.

We bruise him for polluting our air and waters. We beat him for soil erosion. We belt him for forest, range, and wetland deterioration. We excoriate him for slaughtering wild creatures. We knock him for urban sprawl, preempted open spaces, vanishing wilderness. We punch him for landscapes scarred by highways, litter, noise, and blight. We strike him for his not-so-quiet crisis of decreasing beauty and increasing contamination that threatens not only the pursuit of happiness but life itself. We flagellate him again for increasing himself. We yell at him to get on with a quest for environmental quality.

Inflict these punishments whenever the urge comes upon you and you'll be praised, because criticizing Man is the "in" thing to do. Come to his rescue, and you're some kind of a nut. Well, Man old boy, stand up. Dust yourself off. These lines are not to bruise, beat, belt, or flay you. In this month of Thanksgiving, we salute you.

Thank you, Man, for giving us, in your wisdom, a generous— and continuing—supply of game to shoot and to look at, and fish to catch and brag about. Thank you for protecting forests against fire, and for assuring a continuing supply of trees for all kinds of uses, economic and recreational. Thanks, too, for more hours of leisure than our species has ever known and for more ways to enjoy them. Thanks for those roads that whisk us in our pursuit of happiness. And even for all of those gadgets and gimmicks that you come up with—from pocket binoculars and mosquito dope to spinning rods and incredibly effective shotgun shells. Thank you, Man, for laws and regulations that keep us in bounds. And for hearing and acting upon the quiet voice of reason midst loud cries of emotion.

Thanks for an increasing dedication to our culture and to our legacy, so that our children, on a warm spring day, as Professor Hugh Iltis has written, "can feel peace in a sea of grass, watch a bee visit a shooting star, hear a sandpiper call in the sky, and marvel at the incomprehensible symphony of life." The basic and applied research that gushes from your inquisitive mind in a giant spring of scientific progress—there just aren't words for the gratitude here. There's a deep "thank you," too, for challenging Mother Nature when she deserves to be challenged, as well as for a growing appreciation of our ecological underpinnings when Mother Nature needs to be supported. Above all, thank you, Man, for being the first species that cares about other species. As Aldo Leopold once wrote, for one species to be concerned about another is a new thing under the sun. The Cro-Magnon who slew the last mammoth had no compassion. But you, modern Man, mourn the loss of pigeons, and swans, and ladyslippers, and you try to postpone the funerals. In this fact, said Mr. Leopold, rather than in Mr. DuPont's nylons or Mr. Bush's bombs, lies objective evidence of our superiority over the beasts.

For all of this, Man, we're grateful. So stand up this Thanksgiving month, you enigmatic, paradoxical creature. Dust yourself off. Swell those bruised, beaten, belted, and flayed ribs with pride. You're quite a guy after all—even though your days on a seething earth may be numbered.

December

Christmas Books for Sportsmen

Once upon a time there was a little boy who moved to a church parsonage in a strange Wisconsin town. Being a preacher's kid, no self-respecting native youngsters would voluntarily make his acquaintance. So he spent the summer sitting on the inevitable swing on the inevitable front porch, reading and reading from the tall stacks of old magazines that depression-ridden parishioners had contributed to the pastor in lieu of money. As chance would have it, the magazines included the back files of such old stand-bys as *National Sportsmen, Hunting and Fishing, Outdoors,* and *Outdoorsman*—publications that have since gone the way of the passenger pigeon. In their pages in those days were the beginning works of a struggling young outdoor writer by the name of Gordon MacQuarrie.

Technically, Gordon was the managing editor of the Superior (Wisconsin) *Evening Telegram.* Actually he was first and last a duck hunter and a superb writer. His arrival on the outdoor writing scene—and in the life of a young boy—was nothing short of electric. With his light humor, story sense, descriptive ability, and conservation conscience, MacQuarrie made the outdoors come alive with feeling and flair. I can still recite by heart—for I was that little boy on the porch swing—a passage from Gordon's "Galloping Goldeneyes": "When the wild, free things of duck hunting are abroad in the very wind with the storm . . . the means are greater than the end, and every hunter knows it." Gordon was to set the pace in outdoor magazine writing for many years, and to create for the Milwaukee *Journal* the first outdoor section of literary quality and ecological substance. Gordon is dead now, but his writing lives on in the pages of a new book by Stackpole, a collection of his *Stories of the Old Duck Hunters, and Other Drivel.* If you want to make glad the heart of a sportsman at Christmas, you will put a bit of Gordon MacQuarrie in his stocking.

Along with Gordon's essays, you might put Ernest Swift's new book, *A Conservation Saga,* published by the National Wildlife Federation. Like MacQuarrie, Ernie cut his eyeteeth in the backwoods of northern Wisconsin, and then he came on from his position as a county game warden to become director of the Wisconsin Conservation Department, assistant director of the federal Fish and Wildlife Service, and executive director of the Federation. Ernie Swift devoted his life to developing and applying a philosophy of conservation, which for its profoundness, pragmatism, and passion is probably unrivaled in America today. This is his story. It is not a pretty story, because Ernie always fired from the hip. But it is a story that warrants the attention of every thinking outdoorsman. As Ernie himself said, "If it doesn't keep you awake nights, it certainly should."

Marking a Treaty Anniversary

The Migratory Bird Treaty between the United States and Great Britain was signed in Washington, D.C., on August 16, 1916, and proclaimed on December 8, 1916. The treaty marked the close of a long battle by pioneer conservationists in the states and in Canada to bring migratory song and game birds under uniform protection. It effectively ended the market shooting that was threatening the extinction of many valuable bird species in North America. The treaty was one of the cornerstones upon which the modern U.S. Fish and Wildlife Service and the Canadian Wildlife Services were built. It established the primary authority of the federal governments of both nations to regulate the hunting of migratory game birds and to take needed steps to assure their preservation. The treaty provided full protection for many beneficial nongame species as well.

Try putting yourself back into the environment at the turn of the century. The faltering steps of the states to save waning waterfowl were foundering on lack of uniform regulations among states and between countries. Ducks might be protected from shooting in Wisconsin, for example, only to fall prey to market hunters in Canada or Louisiana. The role of the federal government in such matters was unclear or nonexistent at the time, much less the role of compacts between nations. After all, this was 20 years before NRA, 30 years before the Marshall plan, and 40 years before FAO.

It is not too much to say that the Migratory Bird Treaty changed our world. It not only saved waterfowl from almost certain extinction, it helped pioneer the concept of federal intercession on behalf of natural and human resources across the country, and it helped create a climate for a whole series of international agreements. When you see a "V" of geese cleaving the skies, you might just recall that the birds were the first exponents of one free world.

Remember Pearl Harbor?

Thirty years ago December 7th enemy aircraft came out of
the blue over a tranquil island lagoon where great ships lay in-
cautiously at anchor. The U.S. Navy was always supposed to
remember Pearl Harbor and its lessons. But that memory has
turned out to be surprisingly short, because a U.S. Navy outpost
has been taken under surprise attack again today. The besieged
outpost is in—of all places—the quiet woods of northern Wisconsin. It has been caught napping not by aggressor bombs but
by an intrepid band of Wisconsin conservationists wielding paper and pen.

The Navy sailed into the country around Clam Lake one summer to begin testing the first stage of a communications system
that could eventually involve a vast grid of 6,000 miles of cable
and 240 transmitter stations on 21,000 square miles of 26 counties. Called "Project Sanguine," the primary purpose of the low-
frequency, long-range system is to send doomsday messages to
Polaris submarines stationed in the ocean depths around the
world. To be built by RCA, Sanguine would have something
like a $3 billion price tag. Sanguine might mean the difference
between U.S. survival and defeat in any atomic confrontation,
the admirals think. They picked northern Wisconsin as the site
for Sanguine because of the region's unique electronic capabilities that stem from ancient underground rock formations.

But so far the Navy's radio signals have been drowned out by
the voices of conservationists who see Sanguine as an utter des-
ecration of the northland. At the least, they say, miles and miles
of cables would scar lovely country. At the worst, 35-volt currents could literally render the area uninhabitable. Leading the
bombing run have been two volunteer pilots who formed a
State Committee to Stop Sanguine—Kent Shifferd, a history
teacher at Northland College, Ashland, and Lowell Klessig, a
fellow in rural sociology at The University of Wisconsin, Madison. They have been joined by increasingly heavy artillery—
U.S. Senator Gaylord Nelson; Les Voigt, Secretary of the State
Department of Natural Resources; the John Muir Chapter of the
Sierra Club; and others.

In the face of this growing opposition, the Navy has decided it better do a bit of ecological research, and Secretary of Defense Melvin Laird has quietly put a hold on the multi-millions needed to proceed with Sanguine. So maybe in time the Navy will "remember Clam Lake" as a lesson in how not to barge in where citizens are ready to combat threats to the environment, no matter how sacred the sources.

Two Days of Christmas

It is appropriate, I think, that the question of what Christmas means in our day be asked not only from a pulpit but also in an outdoors book. There are two ways to savor the spirit of Christmas, the city way and the country way.

The massive character of an urban Christmas has all the compulsion of a bulldozer. The colorful advertising campaigns of department stores, in homes and hostelries Christmas trees alight with all their trimmings, the "Adeste Fideles" sung heartily by men and women in chapels and bars—all this constitutes a phenomenon of imposing proportions.

In quiet contrast is the countryman's Christmastide, experienced in "the ringing silence of the swamp," where the stars seem to glisten just beyond arm's reach, and where it is not at all difficult to imagine a weird choir bringing surprising tidings to a cluster of shepherds huddled on a nameless hill long ago.

Whether you respond to the sparkling tinsel of Christmas in the city, or to silent nights in the country, it is difficult to find many civilized human beings whose hearts remain unaffected by the anniversary of Christ's birth, no matter whether they consider it myth or historical fact. Indeed, it is hard to think of that birth either as plain, unadorned history or as sheer legend. For the Christian, that event approximately 1,971 years ago is the point of confluence of a miraculous faith. For the doubter, that legend and what came out of it is still the most extraordinary success story the course of mankind has known. Not one of the great explosive events of recorded time can be compared to the concept of "rendering unto Caesar," if for no other reason that all other revolutions can be traced back to that principle. By becoming subject to two separate allegiances, to a God and to a Caesar, the human person was at the same time sanctified and fissioned.

The conflict between these two elements in man—rights of kings and human rights—has fantastically energized the human adventure, even as it has given little chance for relaxed, animal quiet to the human breed. How can we be oblivious—at least at Christmas time—of the fact that the principle of freedom, sometimes called liberalism, is nothing but a translation in political terms of the religious idea of man as the temporary responsible bearer of divine creativeness?

For some of us, we "get into the Christmas spirit" best on an urban shopping spree or at the traditional office party. Others of us require a twilight hike in the hills, The setting is probably not so crucial as the preservation of the option.

January

How's Our Habitat?

What's happening to the soil, water, forests, wetlands, and wildlife of America?

State Departments of Natural Resources annually take the occasion of turning of the year to render a balance sheet under a title like "Trends in Conservation Problems." Each always reveals a congeries of "unwarranted destructive practices." Take these Wisconsin data, for example:

* In spite of added acres of contour plowing and strip cropping, Wisconsin's precious productive topsoil continues to wash and blow away.
* Pollution of lakes and streams is widespread.
* Improper handling of woodlands is still common.
* The loss of wildlife cover is proceeding relentlessly.

Why are our outdoor resources operating in the red? Philip Hauser, distinguished professor of sociology at Chicago University, believes the chaos afflicting man and land today is the product of the transition we are still experiencing from an agrarian, pre-industrial country to urbanism as a way of life. He says we are enmeshed in conflicts between traditional ideologies and institutions and the harsh realities of the contemporary world.

Another recent Wisconsin publication, a "Planning Blueprint," highlights what has been happening to us in the memories of most living men:

* On New Year's Day 30 years ago in 1940, 1½ million Wisconsin people lived in urban areas. Today a million more—2½ million—live in our cities.

* In 1940 there were less than 900,000 vehicles on our highways, driving less than 6 billion miles a year. Today almost 2 million vehicles drive 20 billion miles on Wisconsin highways.

* Thirty years ago there were no ski slopes operating in the state. Today there are over 50.

* Thirty years ago there were only a handful of water skiers. Today thousands of men, women, and children churn up hundreds of lakes and rivers.

* In 1940 there were less than 230,000 licensees fishing Wisconsin waters. In 1970 there were over a million.

* In 1940 there were less than 98,000 big game hunters. This past fall over 400,000 took to the Wisconsin woods.

* In 1940 the state parks had 1¼ million visitors. This year they will be inundated by more than 6 million.

Hundreds of other striking changes could be listed. And even faster change lies ahead. Is there anything encouraging in the picture? There is.

With new federal and state funds we are beginning to acquire more pieces of outdoor living space. Researchers are stepping up the pace of conservation studies, and engineers are developing practical applications of new findings. New federal, state, regional, and county instrumentalities, through planning and zoning, are beginning to try to ration, husband, redevelop, and preserve our natural resources. Above all, a growing segment of the general public seems increasingly committed to a quest for environmental quality.

Will success come to conservation? The story will be told by the balance sheet on New Year's Day, 1999.

On a Lake in Winter

The city youngster who associates skating with a flooded play-ground or even with a park lagoon is missing one of the great delights of the out-of-doors—exploring the glass-like surface of a lake which has been frozen to perfection. It takes a fairly rare concatenation of natural events to turn a lake into a first-class skating rink. Only every ten years or so do you get the recipe: nights of penetrating cold and utter calm, and absolutely no snow. Given these conditions you can wind up with an expanse of absolutely flawless ice a half a foot thick.

It was so with Rock Lake at Lake Mills one day in January, so I took our girls out to our marsh shack for an introduction to real skating.

First we got a fire going in the cabin. There is little so satis-fying as a fire in a pot-bellied "Round Oak" stove. Heat has nothing to do with it. An oil or gas furnace is ever so much more efficient, but the click of the thermostat does not convey the same message as does the popping of a red-hot stovepipe or the soft singing of burning oak.

Thawed out and bundled up, we then whisked over the ice for half a mile in any direction, with only a few baby pressure ridges to mar our passage. We watched ice boaters and skate-sailers. We twirled in utter freedom. We inspected the doings of winter fishermen. One was angling for pike with a live perch for bait. He had caught a mess of small perch with grubs that he had retrieved from the center of an old tamarack log split asunder with an ax. Such a man has a real right to fish. We looked into his pail of perch, and it was as if he had somehow mastered the secret of capturing summer and storing it up for winter use. We marveled at the pickerel he pulled out to lie flapping on the ice. A June pike has no special charm, but a January pike takes on a dazzling beauty. He is not just green like the pines, nor gray like stones, nor blue like the sky; he exhibits rarer colors under the low-lying winter sun, as if he were a precious pearl that glints under only certain light con-ditions.

We looked down through the fishing holes into the quiet parlor of the pike, pervaded by a softened light as through a church window. The sandy bottom stretched away like a bright summer beach. Long strands of pickerel weed undulated as if to the strains of some music forever denied to human ears. Little green-eared sunfish swam up sedately, and then twinkled off to all sides as they perceived our eyes upon them. You may be able to have as slick a skating surface within the circumscribed limits of a city rink as you occasionally find on a lake, but you enjoy only half a world. There is no window to that perennial waveless serenity that reigns beneath the ice as in an amber twilight sky. In Thoreau's words, "on a lake in winter heaven is under your feet as well as over your head."

February

Who Wants February?

Of one thing we can be sure about the Roman priest who fashioned the modern calendar: he was a sportsman. We know this because he made February the shortest month. It is in February that the outdoor year is at its nadir.

The hunting season is gone in February. Oh, you can still go after foxes, and this is the highlight of the year for some hunters, but their ranks are thin compared to the armies that sought upland game and deer in the fall. One thing any hunter can do to kill time in February is take down his gun for a thorough cleaning. For some, taking down a gun provides hours of basement pleasure. For others like me, taking down a gun means taking it down to the sporting goods shop where I can be sure it will get put back together again.

The fishing season is yet to come in February. True, the devotees of ice fishing have a big time this month, but those of us who associate fishing with open water and flyrods very rarely venture forth onto frozen lakes. One thing any angler can do to kill time in February is replenish his tackle box. For some, this means tying flies or whittling plugs. For others like me, it means poring over exotic catalogues and mailing off an order. Even the birdwatching and hiking season is at low ebb in February. The hardier souls continue to fare forth in search of horned larks and crowned kinglets. The rest of us go to Audubon Society movies.

Yes, there is nothing so long as a weekend in February, with not even a bowl game on TV to break the excruciating monotony. Some of us may try to sit through "Wonderful World of Golf," but the sight of those green fairways and blue lagoons in California or Caracas is calculated to cause a fan to foam at the mouth. Others of us take our kids to the ski slopes.

What most outdoorsmen are reduced to in February is weather-watching. There is something about some February weather that is elemental in its allure. There is the February blizzard that blasts away, for a day at least, any trace of civilization. Peering through swirling skies, you cannot make out the house next door. Avenues become as stark white and trackless as a prairie trail. Even staunch oaks and evergreens bow in obeisance to the gale. We are left breathless by this annual reminder of nature's primordial power. On the other hand, there is the February thaw that sets feet slopping through rivulets in city gutters and hearts bouncing like kites in the balmy air. Snowbanks disappear. Daffodils may even thrust up with premature enthusiasm where the higher-riding sun bounces off basement walls. Cock cardinals and pheasants can be heard crowing. We strain our ears to catch the first honks of geese returning north.

But mostly, February weather is drab, a half-way station between winter and spring. The outdoorsman makes the best of it, thankful there are only 28 days in which to wish he could hibernate like a bear. The ancients who invented the 12-month calendar placed in February their gala festival of Lupercalia. The whole populace would gather in the town square for a sacrificial feast. Those who ate enough and drank enough were said to be cured of sterility. Maybe the old Romans had something.

Looking Ahead to Fishing

Give a fisherman a February thaw and he begins to dream.

In his mind's eye he rows out in the stillness of the mists of the morning when the lake is like polished onyx. He rows slowly and silently in the hush of the birth of the day, feathering the oars with a twist of his wrists. He thinks how miraculous it is that the boat moves even when he holds the oars poised in mid-air. What a fisherman thinks about is not to be understood by other men. He conjures up the lilypads, the sedge, the coves, the points of land, the snags, the little recesses, and he thinks, "If I were a fish, I would be right there."

Particularly he pictures a little lane between the pads that leads to a sort of hidey-hole against a deep-cut bank. Suddenly he knows that there is a fish there and a good one, just because he knows it. He greases his line and puts a touch of graphite on it so it will slide slickly through the guides. He opens the bend of the hook so that it will work just a trifle better than the manufacturer made it, and he hones the point until it is needle-sharp. Perfection is a matter of trifles and the February fisherman is a perfectionist. He false-casts with his fly rod, and then he places the bass bug with its twitching rubber legs exactly where he wants it, 'way back in the hidey-hole, six inches from the bank. He lets it lie still. Then he twitches the fly line, bringing the bug forward a notch, and then lets it rest again.

"Hit it! Hit it! Hit it!", he pleads.

Nothing happens. He snips off the bass bug and ties on a split-tailed streamer. The lure is a thing of feathers and it does not look like a frog until the fisherman drops it back into the hidey-hole and gives it action. The fisherman watches it, opening and closing a couple of inches under the water. The black bass sees it, too. He charges out through the little lane in the lily pads and strikes.

With that symphony of coordinated response that only a fisherman can muster, the fisherman sets the hook, desperately pulls in slack, grabs an oar with one hand and gives it a pull, switches the rod, elevates it, and gives a heave to the other oar. It is good luck that the bass is heading for open water. He might have tied up the leader in the lilypad stems.

"Give me room, lots of room, to play him," the fisherman prays. "He is a big one."

The bass comes to the top, half out, and shakes his open mouth and rattles his gills. He is too big to jump all the way out, like a smaller bass.

"Oh, what a bass!", the fisherman thinks. He is a happy man and it is a beautiful morning, even if it is all happening only in a February daydream.

Death Rattle in Dodge City

Back a hundred years ago or so, when I was a boy, every self-respecting farm and even every small-town household kept chickens; and February was the time of the year when we culled the flock. On an appointed Saturday morning the whole family would help single out the surplus roosters and the over-the-hill hens for a mass canning operation.

There is nothing like the crude execution of a Rhode Island Red to bring a small boy face to face with the thin, mysterious line that separates life from death. One moment the fowl is resting quietly on the chopping block. Then the axe descends, and the headless body takes off over the yard in one last paroxysm, a fitful series of gyrations as if in frantic search for what is forever lost, finally to lie utterly still in a patch of red snow.

"Why does it do that?", I can remember asking. My father would say, "It's the death rattle." And I would wonder at the phenomenon of a dramatic flash of activity that signals not a beginning but an end.

There are human institutions, I have since discovered, that recapitulate in their life cycle the decapitation of a chicken. We are witnessing such developments in many Wisconsin counties. My county board, for example, is struggling manfully to enter an urbanized twentieth century. At the moment the old forces of agrarianism would seem to have the upper hand, blocking even any mild attempts at integrated land use planning. Some might think what is at hand is a viable populist revolt against the city, but what we are actually seeing is the death rattle of an era. The spokesmen for township and village enclaves are having a last say, but it will be to no avail. Their world is as dead and gone as the back-yard chickenhouse.

What is at stake, of course, is a practical means of achieving environmental quality conservation, redevelopment, and maintenance in America. The problems of water pollution, air pol-

lution, urban sprawl, traffic snarl, disappearing fields and forests, waning wildlife, increasing litter, noise, and blight do not
confine themselves to township, village, and city lines. They
can't be tackled by geographic bits and pieces, nor can they be
tackled one by one. They fit together in a matrix of people and
practices that will respond only to comprehensive county and
regional planning, as some enlightened members of county
boards are trying to say.

Planning is a sort of dirty word in some quarters; it is said to
smack of regimentation. In other quarters planning is looked
upon as a panacea—some kind of snake oil good for all the social ills of man and beast. The truth is somewhere in between.
Planning is no better than the people who do it and the codes
that enforce it; planning does involve a loss of autonomy on the
part of individuals. But the alternative is even worse—a heedless slide into increasing contamination of the countryside and
decreasing physical and cultural beauty.

The country lies today between two worlds, the one struggling
in its death throes, the other struggling to be born. What we
probably need is a Bart Starr to call some political signals.

The Sounds of Spring

It is quite possible that modern man was never intended to
inhabit an environment like a mid-January day in the north, with
the chill factor hovering around the 75 below mark. The only
way we can make it is to hole up in the jungle hideaways we
have re-created in our homes, warmed to a temperature of 75
degrees above by ancient sunlight stored in fossils. To make sure
these suburban caves are not only physically but psychologically
to our ancestral tastes, we decorate them with the ferns, flowers and grasses of forest, savanna, and veldt, and we invite bird
and animal pets to share the long winter nights with us.

Hairless, and without the traits to hibernate, it is not clear
that we could make it through February even in our humidi-
fied retreats were it not for our divine ability to anticipate the
future and draw inner strength from the visions we thus can
conjure up. Myths about the groundhog to the contrary, man is

the only species who can look ahead, and this may be the single most significant factor in accounting for his temperary ascendancy. He knows for sure, for example, that when winter comes, spring is not far behind.

It helps, though, at this time of year, when we emerge sporadically from our dens, that we hear some of the earliest sounds of returning spring to reassure us. There is, for example, in the first week of February the voice of the cardinal. This does not mean that the cardinal is an early migrant; he has been here all the time. But he is voiceless until a certain light intensity triggers his vocal cords, and then he breaks out in roadside hedges and on city trees with a priceless "what-cheer" that is perhaps the earliest sound of the turning year.

Shortly to follow will be the voice of the ringneck rooster. This immigrant from China has found what they call an ecological niche in the Midwest. Where the grouse finds no sheltering woods, where overgrazing and the plow have forced the prairie chicken out, where the quail has fought a losing battle against long, hungry winters in a stripped countryside—here the pheasant has found some way to live, and after the deepest snows and coldest temperatures his defiant, raucous crowing greets the early spring as he struts across the harrowed fields of farmland and suburban fringe.

A little later will be the spring sound of a river as it throws off its icy shackles with a great grinding and roaring. There is even one sound of spring that we don't have to leave our heated caves to hear. It is that highly distinctive, sharp yet hollow sound of a golf club engaging a golf ball, as we watch the early tournaments on television from Pebble Beach, Palm Springs, and Augusta. Without such intimations of a more hospitable climate on the way, it is by no means certain that anybody would survive to mid-March.

Thoughts on a February Thaw

There is that time of the year when snow and bitter cold hold sway outside, when inside there is not a football game to be seen on TV, and you wonder of a Sunday afternoon, "How will I ever make it?" Then comes that grand reprieve—the annual February thaw. The sun rises with returning warmth one precious morning. By noon gushing rivulets fill every gutter. You go for a sideroad drive. You see things stirring.

A man on a tall ladder is striking the ice from the overloaded eaves of his house. In the woods you see the flash of a lifted axe before you hear the sound of it. A vast flock of starlings cries in a bare elm top. There are hens scratching and cackling in sunny barn doorways. Cattle, long mewed up in dusky stables, yawn complacently in the open yards or scratch their rumpled hides upon convenient posts.

You get out and take a long tramp through waking woodlots and fields. The air is sparkling with welcome warmth. The lingering snow is like a map, with tracings of rabbit runs, the footprints of mice and pheasants, here and there a galloping fox, and often the proof that a squirrel has ventured out. You find many little holes dug through the snow to the leaves and the berries and the roots beneath. The tag alder, shaken by the wind, scatters from its small brown cones a shower of seeds upon the snow—manna for wild, shy birds and shrews. You walk through an old orchard, looking narrowly at every tree for swelling buds. You look hopefully in the duff where you know are hepaticas. With a longing you cannot describe you follow the flight of crows high across the barren fields. You see a flock of juncos among the bushes, flitting about with a kind of lively, fearless excitement, and you know why it was that St. Francis could so love the larks of Assissi that he could petition the emperor for their protection.

There is nothing sensational about a hike in a February thaw, but then the chief part of every life consists of small things anyway. It is strange how competently, even grandly, some men will ride out the great storms of sorrow and tragedy, who are wrecked upon the little reefs that litter the calm waters of their daily lives. Fortunate is the man who can draw from a February walk the assurance that even as nations go wrong, in the country places the unobserved world goes onward with its daily affairs: winter wears to its end, spring comes on, life turns ineffably upon the poles of its ancient routine.

Back home, buoyed up by February sunshine, you take your

shovel and turn a clod of soil, breaking the glebe up with the flat of the blade and spreading it thin on the sidewalk. Out of the mass some white grubs stir; numb beetles on their backs stretch stiff legs; and a chill earthworm squirms his coral length. The split half of a crocus bulb lies there too, slipping out of its old brown coat and showing clear, young white beneath, with just a tip of sprouting green.

And so you know, though wintry days may yet be in store, spring is on the way as the ecological year surely turns.

Part II
OUR PLANETARY PARTNERS

3. Wonderful World of Fishing

I SEE BY THE PAPERS WHERE MORE FISHING LICENSES WILL BE SOLD this year than ever before in the history of the United States. I don't know of anything more than this statistic that so testifies to something tenacious in the human spirit. Because it takes a peculiar, pesky soul to be a fisherman.

First of all, come drought or high water, death or taxes, recalcitrant wives or pleading mistresses, you positively have to participate in opening-day festivities. There is a piscatorial law that says so. The opening hour is not a respectable 12 noon, as in the case with duck hunting. It is dawn. The night before you are always invited to somebody's cocktail party. When you attempt to leave before 1 A.M., people look at you askance and whisper behind your back.

You get up at 3:30, stumble around the kitchen in hip boots, and fumble around in the icebox for sandwich makings. Somebody yells something from a bedroom. The streets of the city are so deserted they are absolutely unrecognizable. You search furtively for the avenue leading out of town. A squad car looks you over suspiciously, and trails you for two blocks. As you near your chosen spot, however, the darkness and solitude disappear. Campfires glisten every couple of yards along the bank. Scores of shadowy figures have preceded you.

You lace on a nightcrawler and cast. The reel you had so carefully overhauled conks out. You work for half an hour untangling the snarl. You cast again. Nothing happens. You switch

back to nightcrawlers. The futile ritual is repeated regularly for
an hour. The sun rises but the fish don't.

At this juncture a nattily attired, wide-awake latecomer el-
bows in at your side. You have to make room for him because
he is Dean Robert Hougas, one of your bosses. Professor Hougas
proposes to use something exotic called a salmon spawn bag—
a little nylon sack of fish eggs, which he says he has prepared
himself from materials acquired in strange places. He makes one
dainty cast—and reels in a trout. He makes a second cast and
hooks a second trout. He takes four more in rapid succession.
Then he pauses because he sees that you are crying.

"Would you like to trade baits?" he asks.

With a disgusting display of gratitude you take his spawn
sack and give him your angleworm. The results are sensational.
He continues to catch trout and you do not. Fortunately, you
now remember that you have an urgent meeting on the campus.
Trailing clouds of ignominy you depart. On the way back to the
city you think up a likely fish story to tell your family and
friends:

"The water was too muddy and the wind was too strong. Nobody was catching anything."

"It hardly pays to go," they will say. But you will return. You
are a febrile fisherman.

Doing Our Thing

On Spring weekends America's great silent majority gets in its
innings, sans fanfare. Let 50,000 or so rock music fanciers con-
verge on a Woodstock farm for a seance under the stars, and the
newspapers treat the event like a second Sermon on the Mount.
But let 500,000 or so fishing fans disperse themselves over a
whole state for an afternoon of nature worship, and nobody pays
any attention.

Yet the similarity between the two groups is striking, indeed.
The rock fancier knows instinctively that his sport is best pursued out in the wide open spaces, untrammeled by the works of
man. So does the fisherman. The rock fancier fills his ears with
the resounding sound of unharnessed drums. The fishing fan is

never so happy as when the roar of rushing rapids drowns out all ulterior noises. The rock fancier wears a special uniform, characterized by its non-conformity with office garb. So does the fisherman. The rock fancier attracts squadrons of sheriff's deputies. The fisherman is responsible for the creation of a special corps of state wardens.

But there is one big difference between the two groups, which probably accounts for the column inches devoted to one and the monumental lack of coverage to the other. The rock fancier is found in large flocks. The fishing fan seeks solitude. So an ordinary reporter can't find him, much less count him.

If the press covered the opening of the fishing season as it did "E-Day," it would discover another striking attribute about fishermen: they are the first practicing ecologists. That is, they put their money where their mouth is. Fishermen buy a special license in order to practice their profession. They also pay a special federal tax on their equipment. The revenue from these two dragnets goes to support the acquisition and management of public fishing waters.

That was really what the message of E-Day was all about— that all of us have to start dipping into our pocketbooks if we want to halt the march of environmental degradation and launch an attack on environmental rehabilitation. Fishermen have been doing it for a long while. When rock fanciers get the same message, conservation may begin to get off the ground. That's the headline we're all waiting to read.

Fishing Facts and Fables

Whenever you tramp along a trout stream or cruise a lake, chatting with your fellow fishermen, you get exposed to as fine a set of assorted fables as you can find anywhere outside of Grimm's Fairy Tales. Here are some of the fables I have overheard, together with the straight dope as presented by biologists:

FABLE—There aren't any fish in here at all. I haven't had a strike all day.

FACT—A good fish population doesn't necessarily mean lots of bites, and an absence of bites doesn't necessarily mean an absence of fish. Even in the space age, fish have their own ideas what to nibble on and when to risk getting caught. At Escanaba Lake, for example, detailed statistics tell this story: In 1954, anglers caught 1,558 walleyes out of a known population of 4,902. In 1959, anglers could catch only 734 out of a population of 5,593. Obviously, fish still have minds of their own.

FABLE—They didn't really put as many fish in here as it said in the papers.

FACT—For every bit of state fish stocking, a planting receipt is executed and filed. This is a permanent record on which is indicated the species, age, length, weight, number of fish planted, date, water temperature of both tank and waters planted, and legal description of the planting site. The receipt is signed by the truck driver, the local conservation warden, and the local "cooperator," a well-known sportsman-witness.

FABLE—I know this trout is a native because the flesh is pink.

FACT—The color of a trout's flesh depends on what it has been eating for the past six months, not on where it has been eating it. A diet heavy on arthropods will typically produce a pink coloration. Most hatchery trout in Wisconsin are fed a diet containing the chitin substance, so a trout fresh off the dump-truck can be pinker than a trout that has been in the creek all winter.

FABLE—The walleye fishing was good last year because they started planting walleyes again.

FACT—The fine walleye fishing last year was due to the excellent success of the native hatch some time ago. Perch fishing in Lake Mendota tells the same story. Perch fishing was very poor in 1957–58, very good in 1960–61, thanks to a fine 1959 hatch that grew nearly eight inches in two seasons. In warm-water species it is the success of the annual hatch that's the critical factor. Maybe in time we will learn how to stock in years with poor natural classes and stock in larger sizes. This is what research is telling us needs to be done.

The latest example of how changeable a fish population can be comes from Lake Koshkonong. This lake hasn't had a highfin sucker population for many years. Last summer they erupted and thousands of pounds were caught. No one could have or would have predicted such a development.

FABLE—I hear they want to raise the fishing license to $3. That's highway robbery.

FACT—A $2 fishing license is so cheap it's almost ridiculous, a $3 license is a steal, a $5 license would be a bargain, and a $10 license would be an honest value.

Fishing is one of the few sports in which you can participate all year, pay for only once, and take home something besides exercise. It's far cheaper than golf, bowling, skiing, or the movies. The same guy that grumbles about a $3 fishing license thinks nothing of spending that much for some cheer to take along on each fishing trip.

FABLE—There's no sense to a big size-limit and small bag-limit on northern pike.

FACT—Some species are very vulnerable to the angler. The northern pike is one. For example, in Cox Hollow one year, 57 percent of the northern population was caught out in the first month. This incident, added to past findings, suggests more conservative northern pike management for areas with heavy fishing pressure.

FABLE—Your Conservation Department buddies always tell you outdoor writers just where to go.

FACT—I always open the trout season on Vermont Creek. Last year I got a grand total of one trout.

Pollsters and Poets Go Fishing

The lakes and streams of the country are full of fishermen— and pollsters. What used to be the simple pastime of "goin' fishin'" has been exalted in recent years as a phase of something called "outdoor recreation," which in turn has been discovered as a fruitful field of study by a growing phalanx of social scientists. So, if you are accosted on your favorite creek or flowage by somebody making a public use or opinion survey, don't be dismayed. You are about to enter the domain of outdoor statistics.

What the people-watchers are discovering about Izaak Walton's ancient and honorable art just goes to show what any fisherman has always known—that fishing has tremendous appeal for a tremendous number of people. Here are some typical statistics:

The magnitude of fish harvested by angling is on the order of one and one-half billion pounds annually. While the nation's population will about double by the year 2,000, the number of anglers will probably triple and the number of fishing days may quadruple. When outdoor fans are asked spontaneously what "usual" leisure activity they engage in "quite a lot," fishing leads the list. Sport fishing was a primary purpose of 4,100,000 visits to 209 national wildlife refuges in 1967—even though most of these refuges were set up to harbor game. Some 15,000 fishermen took 6,204 legal-size muskies from New York's Chautauqua Lake in 1967. Nearly 50,000 anglers caught over 228,000 fish in 216,000 hours in Pool 13 of the Mississippi River. Some 132,000 anglers took nearly 323,000 rainbow from Oregon's Diamond Lake. But provision for more public fishing waters is not keeping pace with the demand. Only four percent of the nation's Land and Water Conservation Act funds went for fishing facilities last year.

With data like these, the economists and sociologists are able to put their finger on certain aspects of fishing, but they will likely have trouble explaining what the sport is really all about, because its appeal is as much in the setting as in the action, particularly on opening weekend. To get the true flavor of fishing we may have to continue to rely on poets.

On T. S. Eliot, for instance, speaking of a month that "frees lilacs out of dead land, mixing memory and desire, stirring dull roots with spring rain." On a current Saturday Evening Post commentator, describing "a sudden apparition of yellow, white, and purple crocuses sprung out of the earth violently and so bright they hurt the eyes; inside of one of these blossoms a bee was rolling about in the pollen and performing a micro-rite of spring." On Edgar Guest, who once wrote in his inevitable way that "The brotherhood of rod and line and sky and stream is mighty fine." Or even on Herbert Hoover, who was not otherwise given to profound statements, but who once said that "Before a fish, all men are equal."

Statistics or not, it could be that the great American rite of "goin' fishin'" is somehow our best single avenue to communion with nature and between men.

What Are We Doing to the Rainbow?

As near as I can place the date, it would have been the first Sunday in June, 1927. My father was delivering the Rally Day sermon at the Barneveld Congregational Church, and his eight-year-old son was required to go along.

On this particular day, however, I was not parked in a rear pew. Instead, I was deposited alone at the rude stone bridge where a primitive County Trunk T passed over Trout Creek just north of town. I had a cane pole, a hank of line, a cork bobber, a Carlisle hook, a can of worms, and unlimited hope. It was not misplaced. Before the echoes of the opening hymn died away, the first trout of my life lay flopping on the bank. It was a rainbow, his gaudy pink stripe and inky back testifying that he, too, had been born in Iowa County.

I would like to recount how I remember the fight he put up, or his record weight. I can't. He was a little fish, and he succumbed to some very crude tactics. What I do remember very clearly of that day is the trimmings: the soft conversation of the stream, the song sparrows speaking from pasture tussocks, the breeze in the box elders, the snort of an inquisitive steer, the passage of a very occasional touring car on the gravel road, and the appearance of only one other fisherman, who gave me a wide berth according to the custom of the day.

It is so with every angler. His standard of quality for an outdoor experience is established very early in his life, and his fishing career from then on is a perpetual search for recapitulation. For many of us, quality is associated with little country rainbows.

On the trail of a memory, I hied myself last opening day to that sacred spot, Trout Creek just below Barneveld, in the company of two other fellow knights of the grail, Professors Robert Hughes and Robert Hougas. It is a mastery of understatement to say that the setting was not that of 40 years ago. In place of a meandering brook there is a man-made lake. In place of a pasture there is a black-topped parking lot. In place of song sparrows there are transistor radios. In place of polled Herefords there are people—by actual counting, 369 of them-elbow-to-elbow around the pond. One of them in particular caught our eye. He was an eight-year-old by the name of Jimmie, on his first expedition. He was wielding sophisticated spinning tackle with aplomb. And he was catching fish. Not just one little rainbow, but a limit of big ones. It did not seem to bother Jimmie that his trout had been delivered by tank truck a few weeks before, or that he had the problem of keeping his line from getting tangled with a dozen others. He was having a big outdoor experience, Model 1970.

I will go no more back to Trout Creek at Barneveld. For me, it no longer offers a quality outdoor experience. But Jimmie will undoubtedly go back many times to seek its big, urbanized rainbows. He and they are all part of a production-line age that has its own measure of excellence. Without question, with our bulldozers and antibiotics we are manufacturing rainbow fishing today in surprising quantity. But perhaps some of us will be pardoned if we still go looking for nothing more than a boy, a bobber, and a brook.

A Day On a Concord Pond

I opened the trout fishing season one year at a small lake called Walden Pond just outside of a little town called Concord, in New England. That is, I appeared on the banks of the pond at the appointed hour. I elbowed in between fellow anglers. And I watched as a few rainbows were horsed in, courtesy of the Massachusetts Fish and Game Department. But somehow I couldn't bring myself to cast a line. It would have been sort of like shooting a mourning dove off a church steeple on Sunday.

Walden Pond, you see, isn't an ordinary trout pond. Or at least it shouldn't be just another pond. It should be set aside as a national shrine. And the fact that is isn't says something rather depressing about the American character.

For those of you who don't remember, Walden Pond is the place where Henry Thoreau built a cabin and explored the universe in the 1840s. What he discovered about man-land relationships has illuminated American thought ever since. In a new book on *The American Environment,* historian Roderick Nash says of Thoreau that "his philosophy became the intellectual foundation of the American conservation movement."

You'd never know it by looking at Walden Pond today. In a nutshell, we've made a mess of Walden. A new four-lane highway slashes across the meadows nearby. At the pond itself, there are the inevitable refreshment stands and bathing facilities. But you can't swim during much of the summer because the pond is polluted by leachings from an adjacent dump. A Niki battery is located just over the hill, and a housing development is pressing in from the west. With some kinds of national monuments we tend to do a better job. In nearby Lexington, for example, where was fired the shot heard 'round the world, the battlefield is pretty well protected and restored, thanks to the federal government. But Walden, a Thermopylae of another kind, is the victim of local exploitation.

It is interesting to speculate what Henry himself would do about such a mis-development. He would probably give a stinging lecture at Cambridge or maybe even over television. He would certainly write a book. He would escape frequently to western wilderness. He might even apply for a Ford Foundation grant. And he would probably go to jail rather than pay taxes to support the steady degradation of his homeland. But he would not organize a Concerned Concord Citizens Association, much less try to get elected to the Middlesex County Board. Because the same curmudgeon spirit that took Thoreau to Walden in the first place left him incapable of participating in the machinery of democracy. Thus Henry may represent a fatal flaw in the American character. Those with a sophisticated ecological conscience are often utterly supine when it comes to catalyzing the kinds of action it takes to save Walden Ponds. And those who get their kicks out of ravishing the landscape know not what they do.

Bass Fishing Is Here To Stay

On the surface, the world of the largemouth bass fisherman has changed a good deal in recent years. For one thing, bass management ideas have changed, and with them the regulations governing bass fishing. Time was when you had to wait until June 20th to fish for largemouths, you could only keep ten-inchers, and the bag limit was small—all in the name of conserving what was thought to be a dwindling supply against the ravages of overfishing. Today the fisheries experts think they know better. They figure that in the presence of acceptable habitat, prolific bass will produce enough offspring to withstand even the heaviest fishing pressure, and that what we most usually have to guard against is too many little stunted bass rather than too few lunkers. So opening day is earlier, size limits are out, and bag limits are bigger. For another thing, bass fishing tackle has changed. Time was when the standard weapon was a ponderous bait rod used to toss a pork rind, a skinner spoon, a harnessed nightcrawler, or a live frog. Today the wand-like spinning rod or spin-casting rod are undoubtedly the most popular and most effective bass fishing tools, and a variety of salt-water jigs have been modified for deep-running inland use as lures: spoon plugs, jig-a-doos, bucktail wobblers, daredevils, bombers, and silver minnows.

But the really important things in the world of the bass fisherman haven't changed at all. July is still the most active month for largemouths. At dusk or just after dark, when faint night breezes condition the tepid air, high-riding lures are still devastating in their effectiveness. A bass churning up from the water to take a surface bait is still an unsurpassed fishing thrill. And there is still that fishing partner who goes along to help—and harass—you with his advice.

Professors Robert Hughes and I were afloat on Rock Lake marsh the other evening to field-test some new artificial frogs. Rowing the boat and coaching was Professor Henry Darling. We pupils weren't doing very well. A half-hour of casting had produced only one half-hearted strike.

"Now let me go over this again," said Professor Darling. "Cast softly into a break in the lily pads. Let your bait rest for a count of eight. Then twitch it ever so slightly, and let it rest for another count of eight. Repeat the process as you retrieve slowly.

"You see that open spot over there? Let's try it once more by the numbers."

Bob Hughes managed to execute a cast according to Coach

Darling's directions, and he let his green frog ride the surface idly. The rings spread out from the bait, the water returned to its boring calm, and nothing happened. Then, as if on a cue from Coach Darling, a big largemouth exploded like a grenade, took the frog in his flaring jaws, shook it like a dog with a bone, and bore down beneath the surface to weave and lash among the rushes. There are other heart-stopping experiences in the out-of-doors: the roar of a grouse in the winter woods, the rip of canvasback over a blind, the snort of a buck in a pine forest. But nothing really can compare with the savage charge of a largemouth attacking a surface lure.

I would like to report that we boated the bass. We didn't. That's another thing that hasn't changed about bass fishing. You lose as many as you catch. It's probably just as well. There's always a bass left to lure you back to the lake.

To Catch a Musky

Every successful musky fisherman has his own formula for success. It may be anything from using chipmunks for bait to taking along his wife's cousin Oliver.

How are muskies actually caught, anyway?

One year, biologist John Klingbiel asked 170 resorts in northwestern Wisconsin to keep records of the muskies caught by their guests. In all, he gathered data on 4,422 muskies from 56 lakes and 6 rivers. For the record, they averaged slightly over 34 inches long. Only ten percent were longer than 40 inches. The average size of the fish didn't vary much from month to month. Over 300 came out of one lake.

More muskies were caught in June and July than in any other months. August and September followed. These figures do not mean the best angling is necessarily in the summer, but only that more people fish at that time of year. What kind of bait is best? Three types took over 75 percent of the fish in John Klingbiel's sample. These were surface plugs, 30 percent; bucktail-type baits, 27 percent; and live bait, 18 percent. All the other types of bait were relatively unimportant. The percentage of fish caught by different types of baits varied greatly from month to month.

Bucktail types were predominant in the beginning of the season, particularly during June. Surface plugs took the greatest number in July and August. In October and November, live bait, mostly suckers, were by far the most effective. These figures, of course, may not compare the effectiveness of different types of lures so much as they indicate the relative popularity of baits among fishermen.

One important point: Very few muskies were taken on baits that are not designed specifically for that species. This means that the angler who fishes for walleyes and bass with normal size spinning or light bait-casting lures has much less chance of catching a musky than if he would use bait specially designed for the trophy species. For pan fishermen, the data are even less encouraging: only five muskies were reported taken on worms. Of course this may mean that worm fishermen are something less than honest in filling out questionnaires.

All in all, Klingbiel says his study adds up to "the fact that musky fishing is a special kind of skill." Although the 12-year-old who lives around the corner may catch an eight-pounder on a perch that had hit on his worm, it is still the dyed-in-the-wool musky fishermen who account for the bulk of this famous fish.

What's the Score?

What happens when so-called stream improvement devices are applied to a trout stream? Do such things as bank cover and current deflectors increase the number of fish? You bet they do. Data from an experimental stretch of Lawrence Creek in central Wisconsin indicate that the average number of legal-size brook trout increased by an even 100 percent after stream treatment. The studies have been carried on over the past six years under biologist Robert L. Hunt of the Wisconsin Conservation Division.

What happens when skindivers invade a lake with their spears? Do they decimate the fish population? Not so you could notice it. Data from an experiment on Nebish Lake in Vilas County indicate that spearfishermen are no more lethal than hook-and-line anglers, if that. Conducting the three-year study was biologist James J. Kempinger of the Wisconsin Conservation Division.

When a lakeshore community sprays its elms, what happens to the DDT? Does it decompose or disappear? Unfortunately, no. It runs into the lake. Three days after the village of Maple Bluff treated its trees one spring, rain runoff into Lake Mendota carried up to 225 parts per billion of DDT, many times the strength required to kill fish. Paul Degurese of the Wisconsin Conservation Division did the measuring.

What has turned suburban garages into drydocks? The boat trailer. The Outboard Boating Club of America estimates there are more than 3.5 billion boat trailers whizzing along the nation's highways this summer, and sales this year will top 175,000 units. Pleasure boating is rapidly replacing the traditional Sunday afternoon drive in the family car as a favorite leisure-time activity, says the OBC.

Will Buck Rogers come to the aid of messy lakes? It looks like he may. Weed cutters have long been used for aquatic vegetation control, but they tend to be time-consuming, inefficient, and expensive. New tools and ideas suggest, however, that mechanical means may yet become an efficient method of weed control, reports Leon Johnson of the Wisconsin Department of Natural Resources. Ultra sound is a new concept. The sound waves erode away the plant cells. Also on the drawing boards are carbon dioxide lasers, argon lasers, and other chemical lasers that build up heat and vaporize plant tissue.

Modern Trends in Fish Management

Good fishing just doesn't happen any more. It is developed by competent fishery biologists.

An example of a modern trend in fish management is the refinement of regulations. Prior to 1945, long closed seasons, exorbitant size limits, small bag limits, and mammoth stocking enterprises were the national vogue. Then came a stage in which all stocking and all restrictions on the take were pooh-poohed. Today we are developing custom management practices for particular species in particular situations.

Wisconsin, for example, had washed out season, size, and bag limits on pan fish, but it has reinstalled stringent regulations on northern pike in southern waters. Another example of a modern trend in fish management is the extensive rehabilitation of un-balanced lakes by chemical means. All states now make use of fish toxicants to poison out stunted populations, followed by re-stocking. Stewart Lake at Mount Horeb is a good example. Once the haven of bluegill runts, it now supports good trout fishing. These management practices were not developed out of the blue. They grew out of a prior accumulation of knowledge through re-search. Just so, the fishery practices of tomorrow will hinge on the basic research of today. The trouble is, as the Sport Fishing Institute says, a great deal of activity now glibly called fish re-search is in reality routine management, akin to checking inven-tory in a supermarket.

Wisconsin, fortunately, has not fallen into this trap, thanks to the practice of turning over $50,000 of Natural Resources De-partment money annually to The University of Wisconsin. Imme-diately practical results or short-term applied research is not sought with these funds. The objective is simply the broad acqui-sition of new knowledge about fish life, particularly in such areas as detailed life history, ecological studies, and behavior studies.

"What do fishes do from day to day, hour to hour, and what are the factors influencing their activities? These are fruitful areas of study," says the Sport Fishing Institute. And these are the very topics now being examined by UW zoologists in cooperation with the Natural Resources Department. Current projects in-clude the biology of trout and bass in artificially alkalized bog lakes, the life history of the white bass, movements of perch in Lake Mendota, and population dynamics of yellow walleye fin-gerlings in lakes.

The white bass study under Professor Arthur D. Hasler has turned up the interesting fact that white bass captured on their spawning grounds on the north shore of Mendota were able to find their way "home," even when they were released at other spots in the lake. Later studies under controlled conditions indi-cate that such fish have a built-in sun-compass that permits them to navigate much as does a mariner. Dr. Hasler hopes to test this theory on migrating salmon at sea in an effort to explain how these fish stay oriented over miles of ocean water. His earlier studies have already suggested that salmon use the sense of smell in order to return to the exact river in which they were spawned.

So the habits of a Lake Mendota white bass may one day explain the habits of a Pacific salmon. In the meantime some basic study of deep-sea biology may shed new light on Wisconsin musky management. This is the way basic research works, and this is the most modern trend in fish management.

The Coho—Boom or Bust?

The coho, sometimes referred to as the silver salmon, is an interesting fish. His native range is the Pacific coast. He is anadromous. That is, he is born in fresh water, goes to sea, and comes back to his home stream two or three years later to reproduce and die. He goes away a five-inch smolt and comes back a fifteen-pound jack. While some anadromous fish require salt water in which to roam, others do not. Along with the sockeye and chinook, the coho can live in a wholly fresh-water environment. Hence his adaptability to the Great Lakes.

Michigan began to stock coho salmon in Lake Michigan a couple of years ago largely to introduce a predator species that would feed on alewives, a little weed fish that has invaded the Great Lakes from the Atlantic Ocean and exploded in numbers in the absence of rainbow and lake trout, the victims of another invader, the sea lamprey. The results of the Michigan stocking were sensational. Fall spawning runs of big coho in the Platte River, Bear Creek, and other streams produced an unprecedented angling bonanza. Fishermen came from all over the midwest to try their hand at catching Pacific salmon. Other results have not been so pleasant. At one spot avaricious sportsmen staged a near-riot in their lust to capture fish by any means; the state police had to be called out to control the mob, protect private property, and preserve the spawning run. For another thing, Michigan now has the problem of getting rid of spent adult fish, which clog streams and depress the commercial market. What is more, the pressure for more and bigger coho is leading to a multi-million-dollar hatchery construction program, because Michigan streams are in no shape to support natural reproduction. Glittering jerry-built marinas are cluttering once-attractive shores, and the economies of country towns have become riveted to the coho craze. It could even be that coho will so decimate the alewife population that the salmon will turn for feed to the fry of desirable species like perch and trout.

Whenever an exotic species is introduced into a new environment, all sorts of things can happen. You can have a friendly alien like the Chinese pheasant or a nuisance like the English sparrow. The coho could go either way. Probably the worst contribution the coho could make would be to blind outdoorsmen to the real problem of the Great Lakes—pervasive pollutions that are slowly rendering these magnificent waters uninhabitable to many wildlife species and even injurious to man.

4. Wetland and Woodlot
Neighbors

TO THOSE OF US BLESSED WITH CABINS IN THE WOODS, SPRING IS
the season of the year for the performing of that delightful
ceremony known as "Opening Up the Shack." We have paid
visits to the cabin throughout the winter, but not for long. An
oak grove in January has a certain austere beauty, but it is essentially inhospitable, remarkably barren of any life. Come May,
however, and a woodland welcoming committee is on hand.

Not all of its members are friendly. One year, when we opened
up the cabin, we found a freshet gushing through the living
room. It was our own fault, actually. We had failed to provide
for proper drainage. Another year our spring was bone dry, the
result of 22 months of below-normal precipitation in Iowa
County. The pond across the road was half its normal circum-
ference. We had to tote our water from another spring down the
pike.

Our woodbox may be inhabited by a family of mice. They
scurry in all directions, squeaking complaints, like the denizens
of any area in the path of an urban renewal development. The
icebox interior is likely covered with orange "gunk"—the result
of a pop bottle that cracked when the temperature hit 13 below.
The eavestrough is plugged where a squirrel piled his discarded
hickory-nut husks.

But most members of our woodland welcoming committee are
friendly. There are the young pines on our perched meadow
sporting a spanking green against the dun grasses. There are the
oaks themselves, living up to their red names by displaying the

109

tiny, wrinkled, dark-pink beginnings of leaves. At their feet are violets, hepaticas, wood anemones, and May apples in profusion.

.   The chairman of the welcoming committee is a ruffed grouse. For six years in a row, he has set up his headquarters on an oak log 30 yards back in the brush from my cabin doorstep in Arena township. Whether it is the same grouse or not, I don't know for sure. The odds are against it, grouse life being pretty hazardous. On the other hand, there are few places safer than a woodlot in which I am the only shooter. I sneaked up on my neighbor one morning as he was performing his immemorial ritual—strutting and drumming to announce that this was his territory and his alone. With a bongo-beating of wings he made the ancient hills echo a signal as old as time.

Few bird and animal instincts are as innate and as prevalent as the territorial instinct. In species after species, an individual or a troop stake out a piece of real estate and call it their own. Interestingly enough, it was not a scientist who helped define the concept of "territory" in birds. It was an Ohio housewife, watching song sparrows through her kitchen window.

Nature, by instilling in the bird or animal a demand for exclusive living space, insures two consequences: first, that at least a minimum number of individuals in any population will be able to breed in relative security; and second, that the surplus will be cast to the literal and figurative wolves who will trim the population to a size the habitat can support.

Those of us who go out into the country to buy a lakeshore lot or a back forty may think we have some very advanced economic and cultural motives for so doing. In reality we are simply reacting to the territorial instinct that our ape ancestors acquired on the African veldt. So my grouse neighbor and I are really very much alike. He will defend with a great flurry of beak and claw his territory against any partridge invasion. I put up big "NO TRESPASSING" signs. He booms the story of his fine territory to all passing females. I put up a cottage as a status symbol.

It will be interesting to see which territory lasts the longest in Arena township, mine or my grouse friend's. The odds favor the grouse. As Thoreau put it, the grouse is sure to thrive, like a true native of the soil, whatever revolutions occur. The grouse has not discovered bombs or biotics with which to defend his sovereignty; in fact, his lease on life as an individual is tenuous. But as a species the grouse is marvelously adapted to the world as he finds it. The same may not be true for us humans.

We are, after all, a comparatively minor and recent species. Our lease in the world is precarious. There is no impossibility, in the coming of time, when a planet loaded with woodlots and grouse may spin forward upon its interstellar journey without a man aboard.

Grouse Hunting the Driftless Terrain

We opened the grouse season in southwestern Wisconsin one Saturday. It wasn't much of a day. The wind made the birds "jumpy." The rain made us wet. The dogs were wild. General Bob Hughes kept to the cow paths while his son and I beat the heavy underbrush. The leaf cover effectively screened what few partridge we put up within range. In truth, we came home emptyhanded. But the stage setting made the hunt exhilarating.

The southwestern Wisconsin stage has been in the making for millions of years. When the rest of the northern United States was being face-lifted by great glaciers, this so-called "driftless" area escaped entirely the avalanche of ice and boulders. Here you can read the signposts of eternity. Sandstone castles and mural escarpments punctuate the southwestern Wisconsin skyline. Springs gush forth from the maws of grotesque crags. Steep hillsides defy grazing cattle and cruising hunters. Hidden valleys provide food and cover for birds.

Most of the area was once sea bottom. Because the hills escaped the glacier's whittling, the rock records of that marine age are still preserved. Showing in bold outline on a cliff back of my Arena cabin are the skeletons of fish with odd armored heads, snail and clam shells, fossil sea weed, and the delicate scrollwork of carboniferous ferns. On the mesas of southwest Wisconsin a lush growth of bluegrass flourishes in the sea-distilled soil, supporting an equally lush growth of beef cattle. But back in the coulees the farming comes hard. Slopes too steep for corn have been allowed to revert to oak-hickory woods, interspersed with

grape tangles, sprawling cedar, and old drumming logs. The cash
crops may be lean here, but the ruffed grouse finds a way to piece
together an existence denied him in more progressive surroundings. This great native American game bird seems to feel particularly at home in southwest Wisconsin, as he jumps with a
roar out of an ancient conifer or wheels silently over the crest of
a razor-back hill to drop out of sight in an eroded canyon.

This is the ancestral home of the Joneses from whom I am de-
scended. None of the Jones boys ever made much money off the
land, but these rugged hills are finally yielding up to a latter-
day Jones an even more valuable crop—partridge and paleontol-
ogy. To cruise these pre-glacial hills and to know that everything
around and underneath has been from prehistoric times as un-
altered as the stars overhead—this gives ballast to minds adrift
on change. You may not always get partridge but you can catch,
in the haze of these mellowing autumn hills, a sense of the peace
that passes understanding.

The Meaning of Hunting

I wind up the grouse season each December, tramping alone
through the Iowa County hills. A solitary partridge hunt is not
an even contest. The odds all favor the grouse. It was so one
Sunday. The score of the game was Grouse, 11—Me, 0.

On such an afternoon you begin to wonder why you hunt at
all. There is, of course, the popular *catchmotive* for hunting provided by parlor psychiatrists: that hunting is a frantic demonstration of masculine virility in an otherwise feminized world.
There is the more sophisticated explanation that hunting sym-
bolizes an urge to escape modern mores and retreat back across
the bridge of millennia to play once more with the artifacts of
our evolutionary youth.

Undoubtedly many of us are still possessed by primordial
spirits. Unquestionably, the freedom of body and spirit encompassed in a hunt enables us to shed certain societal pressures and
walk, for however brief a time, in the open, on ancient quests.
From this perspective, hunting today is the one great basic adventure for millions of ordinary, town-bound men who yearn for

personal participation in some kind of frontier confrontation with the great forces of the universe.

Some non-hunters challenge the basic morality of hunting, questioning the right of a reasoning species to prey on other species for sport. It is true that some hunters are interested solely in the full game bag as a token of manhood and tribal acceptance, but most hunters outgrow this blood urge and come to seek much deeper values. We invest, for example, our quarry with character and worth. To some degree we adapt ourselves to the creatures we hunt, and so acquire a measure of their freedom and sagacity. Particularly, we steep ourselves in the setting of the hunt, in what Charles Nordoff once termed "the spirit of place."

Much more than a lust to kill, the hunter has a rage to live. He may never have heard the word "transcendental," but he senses its meaning. When such a man hunts, it is very difficult to regard his act as an overt offense to the dignity and spirit of game. Rather, it may be a personal testimony to the dignity and value of the wild. The consummate offense in our day to wildlife is not hunting, but the extirpation of habitats by an indifferent technology in which game is wiped out, not by man's passion but by his single-minded devotion to a material world in which wild creatures have no place.

Indeed it is the hunter today, more than anyone else, who feels most deeply man's interdependence with his environment, and who has demonstrated a responsibility for that environment and its maintenance. It is the hunter who will continue to take the lead in cushioning the impact of modern culture on our natural resources, using the most effective political and scientific tools at his disposal. Thus may well be fulfilled Thoreau's prophecy that "in wildness is the preservation of the world."

Can Quail Come Back?

At least once a year we like to leave highly managed hunting grounds and head west to hidden valleys as yet undiscovered by efficient farmers, gunner goons, or eager biologists. Here in the sand country, where the hedgerows grow taller than the corn, where deer tracks are thicker than boot tracks, and where government agencies haven't yet tried to manage much of anything, you could find a native Wisconsin game bird "doing quite well, thank you." He was bouncing Bob White, the quail.

It was not the quantity of quail that made the sand country so satisfying. It was the quality of the hunting. There were no check-stations, no droves of trigger-happy shooters, no artificial targets, no manipulated cover, no posted food patches. There was only a land that has gone to seed, and a little native bird that had been making a go of it since your greatgrandfather broke the Wyoming Valley plain. To be able to harvest a crop of quail from this neglected corner of the world made Bob White more of a trophy than a Horicon goose. Now Bob White is gone from the sand country.

Biologists Cy Kabat and Don Thompson of the Wisconsin Department of Natural Resources have come out with an intriguing proposal called "A Program for Wisconsin Quail Management." In a nutshell, their thinking goes like this:

1. Quail have been gradually declining in Wisconsin for about 100 years. The primary factor affecting this loss is the destruction of the main component of quail habitat: hedgerow cover. If the rate of hedgerow loss continues, quail will disappear in about 10 years.
2. We need to maintain and increase hedgerow cover so that it exists in the ratio of one mile approximately 12 feet in width to 450 acres of farm land in the quail range of the state.
3. To do so will take the combined efforts of the College of Agriculture, the State Soil Committee, the Agricultural Stabilization and Conservation Agency, the Farmers Home Administration, the Soil Conservation Service, and the Conservation Division.

The Kabat-Thompson plan makes a lot of sense, particularly for those broad areas of the country where clean farming has swept Bob White under the carpet.

Pheasant Fiction and Fact

Thanks to an immigrant Chinaman, a lot of hunters are enjoying upland bird shooting on lands that once held native quail, grouse, and prairie chicken, but where now only the exotic ring-neck pheasant can make a go of it. So popular have the pheasant and pheasant hunting become that nearby fields today may more resemble the midway of a county fair than a stretch of country-side. It is probably only natural, then, that there has grown up around the ringneck a great body of fairy stories as remote from the truth as the chant of a carnival barker.

Now, thanks to an exhaustive study of pheasants, pheasant hunters, and pheasant habitat, carried on for seven years in Dodge County, Wisconsin, we can begin to separate fact from fiction. Here are some of the recently reported findings of John Gates, game biologist with the Wisconsin Conservation Division:

FICTION: Pheasant cocks are such good hiders and so fast on their feet that most of them survive the hunting season.

FACT: Wisconsin pheasants are harvested at a higher rate than any other of our game species. Upwards of 75 percent of the cock population is bagged each fall.

FICTION: This means that we are overshooting our pheasants.

FACT: Being polygamous, a pheasant population can tolerate a heavy cock harvest. Dodge County eggs in the wild have an even higher fertility ratio than occurs at the state game farm.

FICTION: That big cock with the long spurs I shot had been around for a long time.

FACT: The odds are against it. About 85 percent of the fall population consists of birds of the year.

FICTION: Pheasants are home bodies; they hang out on the same forty.

FACT: Pheasant populations typically undergo extensive shuf-fling in late fall and early spring. Some individuals are likely to travel up to six miles; the average distance traveled is about one and a half miles.

FICTION: Four- and five-week hunting seasons are bad business.

FACT: Over 90 percent of the season's gun pressure and kill has taken place by the end of the third week of hunting. Length-ening the season beyond three weeks has little overall effect on the pheasant kill.

FICTION: If you don't get your birds the first weekend you might as well quit.

FACT: While it is true that 40 percent of the total hunting

effort and kill take place during opening weekend, hunting success in terms of cocks bagged per hour of hunting declines relatively little as the season advances.

FICTION: You've got to get out there on a weekend when a lot of hunters are in the field to keep the birds stirred up.

FACT: Hunting success on weekdays averages almost half again as high as on weekends.

FICTION: A dog is a nuisance when you're pheasant hunting; the birds won't hold.

FACT: Hunters with dogs average 17 cocks bagged per 100 hours of hunting and only nine cocks crippled and lost from every 100 birds knocked down. In contrast, hunters without dogs average only 12 cocks bagged per 100 hours and lose 16 cripples for every 100 cocks shot.

FICTION: Hunters shoot lots of hens illegally.

FACT: Even though the Dodge County study area sustained heavy hunting pressure, only about 12 percent of the hen population died during the hunting season.

FICTION: The place to find pheasants is in a cornfield.

FACT: As the hunting season progresses, increasing numbers of cocks move as far as four miles from uplands into the protective cover of wetlands.

FICTION: Anybody can shoot pheasants successfully.

FACT: The better equipped, the more experienced, and the better acquainted he is with the area, the more pheasants the hunter bags.

Ringnecks on the Skids

The ringneck pheasant was first introduced into Illinois 75 years ago. It rather quickly established thriving populations in northern and east-central Illinois, but it has never been able to make it in the southern counties below the 39th parallel. Ron Labisky of the Illinois Natural History Survey has just completed a rather exhaustive analysis of population data for the past ten years. He has found that:

1. State-wide, pheasant numbers in Illinois have declined about 44 percent since 1963.

2. In the best pheasant range the decline has been even sharper—60 percent.

3. In contrast, a severalfold increase in pheasant abundance has occurred in a few counties in marginal range in the past five years.

Why the general decline in ringneck pheasants? A change in farming practices, says Labisky. A big increase in row crops like corn and soy beans and a big decline in acreage of tame hay and small grains has meant a wholesale wiping out of pheasant nesting habitat. In other words, it is the farmer and his implements and not foxes or stocking that control pheasant populations.

Labisky and others have also tried to take a look at why the ringneck has never caught on in southern Illinois. The supposition has grown that pheasant range coincides with areas that have been glaciated in relative recent times, and that pheasants can't make it on calcium-poor, geologically old soils. But Labisky and his colleagues haven't been able to detect any difference in the calcium content of the feathers from northern-Illinois and southern-Illinois birds. Whatever is operating to restrict the range of the ringneck is a very complex nutritional matter, the biologists conclude.

Pheasants Need Nesting Cover

If the pheasant hunters plying midwest uplands will look around them, they will see this handwriting on the land.

"Pheasants are losing nesting cover. This fact, and not foxes, accounts for their long-term decline. If we are to stand any chance of halting or reversing this trend, new approaches must be used." There are the words of Cyril Kabat and James Hale, the Wisconsin Natural Resources Department's ringneck researchers.

Anybody hunting pheasants from 1939 to 1967 knows the birds have gone through a series of ups and downs. But from the time when pheasants reached their highest population levels in the early 1940s to now, the long-term trend has been gradually downward. You don't need a crystal ball to know why this has happened. All you have to do is look around when you're hunting. Short-range fluctuations were caused by adverse weather; the long-term decline is the direct result of trends in land use. Drainage of wetlands, destruction of other undisturbed cover, and the advancement of hay-cutting dates are the primary factors in the worsening picture.

What can we do about it? First it is necessary to realize that it is the wild-hatched bird that sustains good hunting. Stocking simply can't carry the load. So it is mandatory to maintain and create habitat conditions that produce reasonable pheasant populations. Next it is necessary to recognize what type of habitat is most urgently required. Biologists used to think winter cover was the limiting factor. In more recent years the importance of nesting cover has become apparent.

Pheasant nesting cover is any grassy or weedy vegetation that is tall enough in spring to conceal setting hens. Attractive nesting sites selected by pheasant hens include hayfields, wetlands, fallow fields, and grain. Unfortunately, hayfields are favorite spots. Twenty years ago this problem wasn't too serious because enough hens brought off their broods before haying time. Now farmers cut their hay earlier to make higher-quality forage, so the peak of mowing and incubating coincide, and the nesting loss is beginning to approach 100 percent in hayfields. In some western states, winter grain is the choice nesting cover, but in the midwest we have little of this type, and spring grain usually is too late to be of much value. That leaves wetlands, provided they aren't too woody, and provided they aren't being drained or burned. Finally, it is necessary to acknowledge that the vast majority of pheasants are produced on private land. It isn't possible to buy enough state land. "Therefore," say Kabat and Hale, "the answer has to lie in working out land-use practices and programs which will provide the incentive for private landowners to create and leave enough nesting cover."

A special group of biologists is working on the problem. What are some of the possibilities?

1. Use state money for taking easements on certain types of acres.

2. Overt acquisition of wetlands with federal Pittman-Robertson funds.

3. Watershed development and soil conservation farm plans that make provision for protecting pheasant nesting cover.

4. Use of decoys to attract hens to nest in non-hayfield cover.

5. Development of materials or devices that will repel hens from hayfields.

6. Use of a sound-emitting flushing bar developed in England for hay-mowers.

7. Continued use of sound harvesting regulations that tailor the take to the available supply.

8. Heavy stocking of adult cocks on public hunting grounds to take the pressure off wild birds.

"The prospect for the future is reasonably favorable," say Kabat and Hale, "if we elect to be realists."

Wildlife Is Going Down the Drain

The frightening story of what is happening to wildlife habitat is starkly revealed in figures released periodically by the Wisconsin Conservation Division after an exhaustive wetlands inventory in each county. Wetlands are these small, damp pockets of vegetation that now represent the last remaining stable wildlife cover in much of the country—sloughs, marshes, potholes, fresh pastures, swamps. They are production centers for pheasants and teal, escape areas, resting and feeding spots, and winter cover.

Take Green County, for example, where the Sugar River flows through glacial drift, and where big herds of Brown Swiss make the milk that makes the cheese that makes Wisconsin famous. This is the story:

In 1939, Green County had 15,777 acres of wetlands. Twenty years later, by 1959, only 7,135 acres remained—a net loss of 8,642 acres for a reduction of 54.8 percent. By 1969 one township had no remaining wetlands at all. No township fails to show a loss. Only three townships have losses under 35 percent. For the county at large, another 44 percent of the wetland acreage re-

maining appears to be easily drainable. The Department of Natural Resources has only this phrase to describe what has happened: "An appallingly high loss for such a relatively short time
period."

Little wonder that pheasant hunters who remember "the good
old days before the war" down along the Sugar find the ringnecks
few and far between now, except where they are hand-planted on
the Albany Preserve. And the irony of the situation is that much
of Green County's drainage has been subsidized by the federal
government through Agriculture Conservation (ACP) payments
to farmers.

Green County is not alone. As a result of subsidized farm drainage in the "prairie pothole" area of Minnesota, North Dakota, and
South Dakota, almost half of that area's 1,350,000 acres of wetlands has been drained in the past ten years to bring to its knees
the principal waterfowl breeding ground in the continental United
States. Little wonder recent wintering grounds counts indicate that
the U.S. waterfowl population has reached a low ebb.

One bright spot is the work of such organizations as Wetlands
for Wildlife, a group of Wisconsin sportsmen organized to raise
money for a wetlands purchase program. The organization started
out by presenting the Wisconsin Department of Natural Resources
with a check for $2,600 to go toward the purchase of land in the
4,500-acre Vernon marsh in Waukehsa County. But it will take a
good many private dimes to catch up with the federal dollars.

Surplus Acres for Sportsmen

HUNTING

permitted without charge

on land in

Cropland Adjustment Program

As you drive through the countryside, look for signs like this
posted along the road. It's a sportsman's welcome mat, put there

through the cooperation of the farmer, the state, and the federal government.

The signs mean that the farmer is participating in an Agricultural Stabilization and Conservation Service program, and has agreed to permit public access to certain acres on his farm on a "no charge" basis for such recreational uses as hunting and hiking. The Cropland Adjustment Program involved seeks—through cash payments from the ASCS to the farmer—to divert land not needed at present for agricultural production into some conservation use. Agreements are for five to ten years. If the farmer agrees to permit public access free, he can receive an extra incentive payment under the program.

The whole program is in keeping with a long-range appraisal of U.S. agricultural and related foreign trade policies submitted by the National Advisory Commission on Food and Fiber. In a nutshell, the Commission report says that the United States has too many crop acres and that some should be shifted on a long-term basis to less intensive uses, such as grazing, forestry, and recreation. "New technology in agriculture is increasing both yields per acre and output per man hour at a much faster rate than the increase in demand for farm products," the Commission said. "This excess manpower and excess crop acres are the heart of the U.S. agricultural problem."

Wildlife agencies, concerned about the long-term outlook for farm game and waterfowl, are finding a good deal of encouragement in the Commission statement that "it is not wise public policy for the government to help pay for draining a farmer's land in order to increase production when we are trying to adjust agricultural production downward." The Commission recommends that public subsidies for capacity-increasing farm practices be discontinued, and that programs designed to switch surplus cropland to grass, forestry, and recreation be "redesigned and expanded." Many of the Commission's suggestions undoubtedly will take form in new legislation designed to combat rural poverty, moderate social readjustments, balance agricultural production and demand, and benefit wildlife in the bargain. Meanwhile, those "Hunting Without Charge" signs on U.S. fields are tangible evidence that the Commission's ideas are already being put into action.

A Letter to Ann Landers

I see by the paper that you take a very dim view of hunting and hunters. In your usual pontifical style you say you don't like "real guns for big boys" because "guns are for killing and I can think of better things to do with time." You are, of course, entitled to your opinion. The only trouble is, if my household is typical, your opinion is taken by the female sex as being only a little less official than the Gospel According to St. Luke. So, in an effort to protect the harried husbands and sons of the country, let me try to offer a few counter-arguments.

In the first place, Ann, the ubiquitous rifle or shotgun over the American mantel has been a most fortunate tradition. From Concord to Veracruz, from Belleau Wood to Bastogne, the homespun American sharpshooter has written history. Profoundly as we might wish it to be otherwise, this country can still use the marksmanship that hunting inspires. The American tradition of hunting has been a fortunate one in another respect. It was the hunters of the country who fostered the conservation movement long before it became the hobby of the League of Women Voters. Wisconsin's Horicon Marsh is a good example. This great preserve has saved the geese of the Mississippi flyway from the ditchers, drainers, dammers, and dumpers. Horicon was earmarked by a hunter hobby, bought by hunters, and paid for by hunters. It is so with many of our other outdoor resources that are only now entering the ken of the environmental crusade. In the third place, Ann, where hunting regulations are enlightened and enforcement adequate, human hunters do nothing more than crop the annual surplus of game that would otherwise fall prey to natural predators or foul weather. You cannot stockpile game birds or animals. You can only select the cause of mortality.

Now I would be something less than accurate if I implied that hunters are always motivated primarily by marksmanship, or conservation, or humane cropping. So let me close with the sort of emotional argument with which you are most familiar. Hunting is essentially a form of aesthetic experience: the first rays of the dawn sun entering a squirrel lot to touch with gold the tops of the hickories; a mallard hen pirouetting down out of a gusty sky like an autumn leaf, to hang suspended for one awful moment over the decoys; the silvered wings and burnished breast of a cock pheasant rocketing up against a dun background of frost-seared corn and sedge; the heart-stopping roar of a grouse as he breaks the stillness of the winter woods; a buck deer sneaking like a rusty ghost through a stand of sighing pines.

This is the world of the hunter. It is what he seeks as an antidote for asphalt and neon. Hunting is a passport to yesterday, a down-payment on tomorrow. It is a remarkably effective cure for urban ulcers. It is, if you please, a brief escape from the cares of your column. As you yourself might put it, "Take your boy hunting and you won't have to go hunting for your boy."

Gun Control Is a "Must"

It is time, probably long past time, for us sportsmen to reappraise our position on gun control laws. In keeping with hunters and shooters all over the country, we have been congenitally opposed to any legislation that would inhibit the purchase and ownership of sporting arms. We have thought we had good reasons.

The right of the American citizen to keep and bear arms is implied in the Constitution, an echo of liberties purchased dearly at Concord and Yorktown. The saga of the gun that won the West is part and parcel of American history. The tradition of the fowling piece over the mantel has led directly to a novel American land ethic in which public access to vast areas of domain stands in striking contrast to the closed societies of Europe. The ubiquitous squirrel rifle has sponsored a widespread native marksmanship that the Sergeant Yorks of the country have put to good use.

All of these changes and more have been rung over the years by the spokesmen for the National Rifle Association, the Sporting Arms and Ammunition Institute, and other groups dedicated to untrammeled ownership of weapons. And we have been bemused by their arguments. Now it is time to take another look. The tragic events of the recent past cry out for change. As Senator McCarthy says, the morality of the frontier is no longer applicable. We are not a country of freewheeling pioneers but a country of desperate community stresses. As Senator Kennedy himself said, it is time to put away childish things.

So what if I have to register my Ithaca pump gun or get permission to carry my .22 pistol? So my freedom is thereby infringed. But the infringement of personal freedom is a concomitant of a surviving society. Sure, I know the argument that a gun is nothing but an inanimate object, and that the problem is the people who wield them. I know that gun control legislation, no matter how carefully devised and rigidly enforced, can't possibly prevent all illicit use. But gun control legislation will help. And we need help, desperately. The American political system is more precious than a duck hunt, the safety of public figures more urgent than a stroll in the deer woods, the sanctity of human life more compelling than target practice.

We have had one too many assassinations. We have laid the tools of violence in too many laps. It is time to act.

5. Whitetails and Men

THERE IS QUITE A CROWD AT CAMP RANDALL STADIUM IN MADISON, Wisconsin, each fall Saturday, but it is small compared to the 400,000 Wisconsin citizens of all ages, sexes, and conditions of servitude who will take to the woods the same season in pursuit of some 800,000 Wisconsin deer. That there are plenty of deer to shoot in this second half of the twentieth century in mid-continent America is one of the marvels of nature. That there are plenty of deer shooters left in this day of lavender plumbing is one of the marvels of man.

Nothing is so orderly in a fundamental way as a deer hunt. Cordons of men merely perform the functions of nature formerly reserved for screwworms and cougars. The deer hunter revives, in play, a drama inherent in the life of the wild. In the last analysis, perhaps, a hunt is an aesthetic exercise with a sure purpose that serves both man and beast.

The lure of the deer woods is compounded of many factors: the camaraderie of the camp, the challenge of the drive through a strange swamp, the almost unbearable suspense on a stump to the flank of a well-trod run, the sheer beauty of a sleek gray coat and a white flag disappearing into a copse of aspen, the skill of the well-placed shot. But the special magic that surrounds the relationship of deer and men may very well stem largely from a subconscious realization that here are two species of the animal world that have been pitted against each other from time immemorial and that have carved out success stories together.

Like man, the white-tailed deer is a creature of the edge. That is, he thrives best in the brush where woods and fields join. Plow up his openings and level his forests and the deer population

125

slumps. But give him a nice combination of woodlots and meadows and the deer will rebound with alacrity. Just so we are coming to see that man himself needs edge if he is to flourish. Crowd him into cities or spread him too thinly over the plains and he has problems. But give him environmental variety and he responds with a cultural irruption.

It is in their periods and sites of stress that we see most clearly the kinship of deer and men. Ranging freely in summer over well-balanced terrain, the deer thrives. Jam him together in a winter yard, and he eats himself out of house and home. Ranging freely between city and country, Wisconsin man thrives. Jam him together in an urban inner core with no escape hatch, and he reacts with violence.

Half a century ago, axe, dog, gun, and trap seemed to have spelled the end of the Wisconsin whitetail. Then a combination of accidental habitat changes and purposeful hunting regulations began to produce a situation in which we literally have more deer today than there were in pioneer days. The return of the deer has been accompanied by the return of the deer hunter. Where less than 25,000 took to the deer woods in the 20s, we can confidently look forward to half a million Wisconsin deer hunters in 1980. Other states tell the same story. Man and game have not generally flourished together. The whitetail story is different. Whether man can survive as well is another question.

Confessions of a Protester

I was one of those who joined the ranks of the Protesters last November. Protesters are a sub-species of humans who have as their rationale an escape from organized society. To distinguish themselves from Conformists, all Protesters wear uniforms, consisting of plaid shirts and soiled pants. All Protesters are also unshaven. All Protesters engage in marches, sit-ins, and other disturbances of the natural environment. Actually there are two types of Protesters. One type is typically found in urban settings like a university campus. This type is commonly called the New Left. The other type of Protester is found in rural habitats like a cedar swamp. This type is referred to as the Deer Hunter.

It was the latter Protest Movement that I joined, along with three other fugitives from the city, Professors Robert Hughes, Robert Hougas, and Harland Klagos, plus three junior-grade Protesters. In true Protester fashion, our first tactic was to hold a committee meeting. We chose as our site the Hughes retreat near Mazomanie. We barricaded the access road and called the meeting to order. The first item on the agenda was a discussion of how best to disrupt the Dane County deer herd the next morning. The senior members of the Cell proposed peaceful tactics like sitting-in along various deer trails. As a matter of fact, one of us would have been satisfied with a lie-in at the cabin. But we were shouted down by our junior colleagues, who insisted on a militant march through the hills. We then turned to a more academic subject, Economics 105, described in the catalogue as "Low Hole Card Wild." Here again we senior Protesters were out-maneuvered by the New Generation. Michael Hughes in particular broke up the meeting by pocketing all the legal tender.

A half-hour before dawn the next day found us all heavily engaged in Protesting. With yellow signs on our backs, shouldering weapons of war, and uttering vile curses at blackberry brambles, we violated the peace of the countryside with true escapist abandon. The senior members of the Cell tried to exercise some semblance of control over the goings-on, but the New Generation insisted on taking over. It was not Chairman Hughes who shot the first buck, it was young Michael. At least we got him to agree to tag his animal and check it in at the state station. As we told him, even Protesters have to live by certain rules. Our Cell has now disbanded for a year. Each of us has gone underground, posing as just another conformist. Our plaid shirts are in mothballs. We have even shaved. But come another deer season and we will become Activists once again. That is, provided there isn't a law against Protesting.

Deer Hunters Never Say Die

Ever since the days of James Fenimore Cooper's characters, the deer-slayer has epitomized the American way. Roaming the frontier as wild and free as his quarry, he has represented that rugged individualism that has been the essence of a muscular young democracy. A modern deer hunter is something else again. A deer hunt today has really very little to do with shooting a buck. It is a desperate attempt on the part of urbanized, computerized man to recapture the soul of his primeval past. But the price the deer hunter pays for his quest is staggering. He comes closer today to representing a character out of Orwell's 1984 than he does Daniel Boone. No other American is so hedged in by restrictions as is the modern deer hunter.

First, the deer hunter today is told to buy a special tag, which he then must display in the middle of his back like a human auto license. He is told to buy special clothing, as least 50 percent of which must be red. He is told when he can go—on three to ten precious days in November, between prescribed hours. He is told where he can go—to certain counties or deer management areas. He is told what he can shoot—bucks or does depending on the area; and what he can't shoot—no small game in the deer woods. He is told what he can use—a rifle, a shotgun, or a bow, of certain strength, again depending on the area. He is told how he can hunt—no elevated platforms, no shooting from a road, no carrying an uncased gun in a car. If he is lucky enough to shoot a deer he must register the animal at a state checking station before he can bring it home. The only thing that isn't regulated is the number of hunters allowed in a certain area, and that regulation is just around the corner. Not only must the modern deer hunter thread his way through manifold man-made restrictions; he must also combat the immutable laws of human health. His chair-bound physique is suddenly called upon to exert itself in ways known only to the pioneers. The price is a rash of heart attacks.

According to all the theories of sociology, there should be no more deer hunters. Fear and frustration should long ago have confiscated the sport. No other human activity is so subjected to harassments. Yet the fact remains there are more deer hunters today than there were 40 years ago. There can be only one explanation for the survival of deer hunting. To drive away from the city, to get out of a chrome-trimmed car, to fade into the woods, there to see a gray ghost of yesterday drifting through the brush—this must constitute a primitive, heart-pounding thrill that can surmount any hazard that civilization can place in the way of its search. Far from representing freedom, the modern deer hunter represents regimentation. But his spirit remains unquenchable. In this there may be a profound lesson for anyone who thinks he can change the American character.

The Whitetail Success Story

Of all man's animal colleagues, you would have to say the whitetail deer has made a signal success out of civilization. If you stop to think about such things, it is really quite remarkable that in a day of I-highways, there are vastly more deer than there were 40 years ago. What may be even more remarkable is the fact that in a day of supermarkets, some 400,000 Wisconsin citizens go back in the boondocks annually on the trail of a buck. Given favorable weather, this annual deer harvest can approach 100,000 animals. So flourishing is the Wisconsin herd that a take of that magnitude won't hurt a bit. In 1924 only 7,000 deer were shot, and the season was closed entirely in 1925.

The whitetail is a Wisconsin native, his bones being found in the earliest Indian refuse heaps. It is quite probable that in prehistoric Wisconsin deer ranged over the entire state, but north of what is now Wausau he was not found in great abundance. The virgin pineries of northern Wisconsin were, in fact, not exactly a haven for game of any kind. Pierre-Esprit Radisson, who cruised the Lake Superior shores in 1658, wrote the earliest deer hunter's lament: "It is a strange thing when victuals are wanting, work whole nights and days, lie down on the bare ground, the breech in the water, the fear in the buttocks, the belly empty, the weariness in the bones."

It was in central and southern Wisconsin, where the woods gave way to oak openings and prairie, that the whitetail flourished, for he is a creature of the brushy edges between forest and field. With the coming of the white man, the deer's southern haven in Wisconsin did not last long. Axe, plow, gun, dog, and snare made heavy inroads. When the first State Legislature convened in 1850, the "deer problem" was on the agenda.

Meanwhile the first lumbermen were creating openings in the

north, and the deer population jumped. After the Civil War the
railroads ran "sportsman specials" to Wausau and beyond. But
by the turn of the century massive clean-cutting operations and
devastating fires had wiped out the northern range. When the
first Conservation Commission was appointed in 1927, there
wasn't much in the way of a deer herd to conserve. Then nature
and man began to team up. The northern cut-over became one
vast sweep of saplings, as reforestation and fire prevention programs took hold. In the south, game and farm management practices favored the creation of more edge. The deer bounced back,
slowly at first and then with a rush. Now the sound of hunters'
guns can be heard from the shores of Green Bay to the foothills
of Dane county.

A lot of people take credit, and rightly so, for the current
health of Wisconsin's deer herd. Something should be said, too,
for the whitetail himself. The cougar and the wolf are gone, but
the deer continues to thrive. His wariness, his adaptability, and
above all, his good will, have paid off.

The Care and Cleaning of Savages

Annually there is abroad in the woods and watering spots of the
country a peculiar breed of savage known as the deer hunter.
To all except his whitetail target, he is not particularly dangerous, but it is easy to hurt his feelings and even to provoke
his animosity. So that you who are not deer hunters will know
how to conduct yourselves properly in the company of these
savages, this book offers some suggestions:

To conciliate the savages you must be careful never to make
them wait for you in embarking.

You must provide yourself with a tinder box or a burning
mirror, or with both, to furnish them fire in the daytime to
light their pipes, and in the evening when they have to en-
camp. These little services win their hearts.

You should try to eat their salmagundi in the way they pre-
pare it, although it may be dirty, half-cooked, and very taste-
less. As to the other numerous things that may be unpleasant,

they must be endured for the love of God, without saying anything or appearing to notice them.

It is well at first to take everything they offer, although you
may not be able to eat it. The barbarians eat only at sunrise
and sunset when they are on their journeys.

You must be prompt in embarking and disembarking.

You must so conduct yourself as not to be at all troublesome to even one of the savages.

It is not well to ask many questions, nor should you yield
to your desire to learn the language and to make observations.
This may be carried too far. Silence is a good equipment.

You must bear with their imperfections without saying a
word; yes, even without seeming to notice them. Even if it be
necessary to criticize anything, it must be done modestly and
with words and signs that evince love and not aversion. In
short, you must try to be, and to appear always, cheerful.

Now these aren't just my suggestions on the care and cleaning
of deer hunters. They are words of an expert, Father Jean de
Brebeuf, in *The Jesuit Relations of 1637:* "Instructions for the
Fathers of Our Society Who Shall be Sent to the Hurons in New
France." Voyageur Brebeuf would feel right at home in the
woods today, and he wouldn't have to change his code of conduct one bit.

The Lament of a Lousy Hunter

If you are frequently invited to accompany assorted colleagues
on assorted hunting expeditions, it may be because of your good
looks or your good spirits. On the other hand, if you are like me,
it will be because you are a poor shot. A lousy marksman is a
handy gadget to have along on a hunt. He can perform menial
tasks like brush-beating and dish-washing, he can be positioned
in a marginal spot without risking the efficiency of the expedition, and his poor performance makes everybody else appear
expert by comparison.

It was last year that I began to suspect the role I play as a
hostage to the gods of the chase. The last day of the deer season

Professor Bob McCabe invited me to go deer hunting with him on his farm. I was flattered. I was even more flattered when he put me on a stand and made the first drive himself. No deer came my way. "I didn't expect to push anything out here," Professor McCabe said cryptically. He then placed himself on a likely stand and told me where to drive. I was only half on my swing when his shotgun boomed. By the time I got there, Bob already had the buck dressed out. "Good thing it was my turn to stand," he said. I had the privilege of dragging the deer to the road.

Early the next grouse season I was asked to go along with Professors Robert Ellarson, Robert Hughes, and McCabe on a partridge campaign. For our assault on a wooded draw, I was positioned in the middle of our line in heavy cover. I flushed feathers but I didn't get a bird. They careened off to the left and right to be taken in the open by my partners. When I complained about the situation, they said, "We planned it that way." The real moment of truth came during the pheasant season when I was invited to accompany Professors Robert Hougas, Harland Klagos, General Hughes, and his two sons, Bobby and Kevin. It was Kevin's first hunt, but at 12 years of age he already knew the rules of the game.

"You work through that ditch and I'll take the field," Kevin told me.

I put up a rooster. Kevin and I fired simultaneously.

"Who got him?" Kevin asked.

"You did" said Professor Hughes immediately.

"How do you know?" I asked.

"Because with the track record of you and your 12 gauge, the odds are all in favor of Kevin and his 410."

I cried like a baby.

My morale had lifted a bit by the time I was invited to join the same crew for the opening of the deer season. We assembled at the Hughes CP near Mazomanie. At H-hour my partners prepared to sneak away.

"Where do I go?", I asked.

"You, Colonel," said General Hughes, "will take your car, drive me to my stand, bring my guncase back to the cabin, and fix brunch."

"Can't I even take my gun along in the trunk of the car?"

"Let him take it," said Prof. Hougas. "He won't disturb anything out there on the flank."

For once my partners reckoned wrongly. Tramping through the woods, they saw no deer. Cruising back along the town road, I did. Off to my left I spotted a buck sneaking along the ridge. I dismounted and slipped through the underbrush to position myself in the path of the animal. It was almost as if the deer knew the caliber of his adversary. Fearlessly he came forward and stopped not 60 yards away to stare brazenly at me. I raised my shotgun, took deliberate aim, fired, and scored a direct hit— on a nearby birch tree. With what I guess was a snort of disdain, the buck bounded off, saluting me as he went with his white flag. Somewhere in western Dane county next morning a young buck was telling his harem, "In the case of some hunters, the only thing you have to fear is fear itself." Someday, perhaps, I will recapture my GI shooting eye, and will then be able to hold up my head in the presence of my fellow hunters. The only trouble is, if that time should come they will no longer invite me along.

A Salute to the Real Deer Men

For a special breed of citizen, the deer season is the signal, not for a ten-day outdoor binge, but for an excruciating period of hard work. He is the field person employed by a state conservation department. It isn't just the wardens who police the deer shoot, although they take the lead. Fish, game, forestry, and park personnel are also mustered to the woods to cruise the back roads, man the check-in-stations, and perform all those manifold chores associated with managing the assault of humans on whitetails.

Conservation commissioners and directors and their lieutenants may come and go, but it is the field people who stay to keep the outdoors green and gamey. I was the fortunate guest the other day of one of the real veterans of the woods wars, E. E. Davison, forest ranger in the Crandon, Wisconsin, area. Davy has been a Conservation Department man since the early 1930s. Most of that time he has spent right in Forest County.

Davy has been shot at by poachers, scorched by fires, nipped by blizzards, reviled by local hot-stove-league experts, and harassed by Madison bureaucrats. He has had his chances to move up or out, but he has chosen to stay in his green field uniform —manning watchtowers, grubbing out lanes, reseeding the slopes, subbing as a warden, stocking fish, and always practicing that lifelong learning that is the mark of the professional man. Now he can look around and know that he has helped mightily in bringing his north country back from a scarred moonscape to a land of rich timber and tourism production.

On his own time through the years Davy has been a lumberjack, and in the course of his cruising he has come by a good deal of tax-delinquent property. It is only poetic justice that this land is now worth a small fortune, particularly the frontage and the islands he owns on Lake Lucerne. The reason this once-discarded land is now worth a lot of money is due to a significant degree to the WCD Davisons who have devoted their lives to the day-to-day chores of Wisconsin conservation when the rest of us were just writing or talking. But the real heroes of the story are the wives who keep the home fires banked while their men are in the woods. There are a good many ten-day widows scattered around Wisconsin this month—women like Frances Davison. She has backstopped her husband every step of the way, in season and out.

The Davisons now live in the handsome stone home they have built on a prime overlook on Lucerne. When you peer out their living room window at sunset, you can see the last rays of light filtering through mature spruce and maple and aspen to glisten on blue waters, and you can expect to watch a doe as she comes delicately down for an evening drink. The sight is a far cry from the Forest County aspects of yesterday, when you could drive for miles through fire-blackened cutover. You can see something else through that living room window on Lake Lucerne. You can see what it means to make a success out of life.

The Everon Davison breed is passing. Davy himself will retire in a year or two. No brass band will mark the departure. But the hunter hordes this year might well pause to drink at least one toast in a country tavern to the men—and their women—who make each year's deer season possible.

6. The Geese Save Us

FOR HUNDREDS OF THOUSANDS OF FOLKS, THE GOOD LIFE IS LINKED inextricably with the sights and sounds of wildfowl. Almost any fall or spring night you can hear it—the querulous, clarion call of migrating Canadas. At first it may be so faint and indistinct as to be mistaken for the background static on a distant radio. Then, as the long line whipsaws high overhead, a wild chorus from a hundred straining throats proclaims the turn of the seasons in a song as old as time, yet ever new.

Winging steadily in flying wedges and wavering lines, the leaders calling the tired stragglers on, their eerie chant ringing across the breadth of wilderness and metropolis alike, the Canadas call to us. Listening in the dark we mark these nomads of the night as they journey down the trackless sky trail. Our minds are moved by the old riddle of bird migration; our hearts are stirred by a compelling kinship with nature.

The Canada goose, like perhaps no other outdoor denizen, has had the power to inspire a concern for conservation in human breasts. A generation ago the Wisconsin chapter of the infant Izaak Walton League was formed to restore Horicon Marsh. Slowly the area was converted from a thistle-infested waste to a pulsating slough. And the geese came back. The masses of big birds milling this fall over Dodge County represent more surely than a moon missile a triumph of man. For one species to protect another is really a new thing under the sun, as Aldo Leopold once observed. We have saved geese. In that fact rather than in Commander Shirra's rocket lies objective evidence of our superiority over the beasts.

Tradition has it that geese in the Temple of Juno once saved

the city of Rome. In 390 B.C. the Gauls attacked and drove the
Romans to a steep, rocky hill known as the Capitol, which was
used as a fort. One night the counsel Manilus was awakened by
the crackling of the sacred geese. Rushing to the wall, he saw
that the Gauls had almost climbed it. His shouts and the noise of
the geese alerted other defenders, and Rome was saved.

The sights and sounds of Canada geese in the skies over
America likewise alert us of the twentieth century. There is a
profound message in the music of migrating Canadas. It says,
as Stewart Udall has written, that "our conservation challenge
today is one of quality—purity of surroundings, and opportunity
to stretch, a chance for solitude, for quiet reflection." The message of the geese reminds us as well that Henry Thoreau's decision to "live deliberately"—to absorb the natural world around
him, not merely through the senses into his physical being but
into his deepest thoughts, to scorn artificiality and find richness
in simplicity—is the nutrient of a great culture and a more peaceful world order.

So as we save the geese the geese save us.

The Contrary Canada

The Canada goose is a bird of striking contradictions. He is
very likely the most generally well known of any of our wildfowl, and yet probably only the most avid birdwatchers know
there are at least 11 races of Canada geese, ranging in size from
the three-pound cackling Canada of the Pacific coast to the 18-
pound Canada of the Midwest. The most common race is the
medium-sized Todd's Canada. Canada geese have been more
persistently hunted, over a wider range of country and for a
longer period, than any other American game bird. Yet today,
in the face of a sharp decline in overall waterfowl numbers and
an increase in gunning pressure, the Canada goose is generally
doing quite well, thank you, and may even be present in greater
numbers in some places than in prehistoric times.

For generations the Canada goose has been the epitome of
wildness, evoking with his semi-annual migrations a sure evidence

of far-away places with strange-sounding names, and representing with his aerial strength, grace, and acumen a contempt for the accouterments of civilization. Yet a Nebraska biologist has recently called the Canada "the most easily managed and the most manipulated species of wildlife on the continent." The traditional stories about the Canada describe him as wary and sagacious. Yet Canadas are actually quick to lose their suspicions of man. Flocks can develop a habit of careless dependency. Biologist Art Hawkins has even called the Canada just plain stupid at times.

One of the reasons the Canada goose appeals so strongly to people may well be the widely held understanding that geese mate for life. Actually, adultery, divorce, and remarriage are not unknown among Canadas. As one researcher has said, "We have seen instances of mate-switching that would do Hollywood stars proud." The range of the Canada may still be said grossly to encompass the sweep of the continent, but new evidence strongly suggests that great numbers of birds are increasingly wintering well north of their traditional grounds.

Some biologists think the Canada goose is pretty well understood. Others say what we still don't know about Canadas would fill volumes. Certainly we still don't know much about how to manage goose hunters, around whom most of today's goose conservation problems cluster. The comeback of the Canada goose during the past 20 years is perhaps the greatest success story in wildlife management, but the resulting situation around certain refuges represents one of the darkest hours in American sportsmanship. Despite the contradictions inherent in any look at the Canada goose, however, on one point there is general agreement: He commands our universal affection and respect. He is the big game of our wildfowl—the king, the aristocrat, the trophy species—whether he is pursued with gun, camera, or eye.

Little wonder, then, that a species of waterfowl so ensconced in American hearts should be the subject of increasing attention by a growing battalion of biologists as they seek to conserve, maintain, and develop our wildlife resources. What they have discovered we'll take a look at in a couple of succeeding sections.

How's the Canada Count?

How is the Canada goose doing in this country today, anyway? There are, according to recent estimates, about a million and a quarter Canada geese in North America, give or take a hundred thousand or so. This is an interesting figure, but not a very meaningful one. We can't manage Canada geese in the mass because they aren't homogenous.

One way to come to grips with the situation is to think in terms of goose travel routes, called flyways. There are four main ones. Another way to cut the problem down to size is to recognize that there are at least 11 subspecies of Canada geese. An even more useful approach is to divide the birds into ecological units known as populations. This term is used to designate all components of a large group of birds typically using fairly well-known breeding grounds, migration routes, and wintering grounds. Thus we can speak of the Mississippi Flyway, having within it a Mississippi Valley Population of Canadas, made up in part of the Horicon Flock and composed principally of the subspecies "interior."

The Story of this Mississippi Valley Population in the past 30 years includes one of the brightest chapters in the history of wildlife management—and one of the darkest. On the one hand, we have brought the birds back from a low of 50,000 to an estimated peak of 300,000. On the other hand, we have shot more geese in one short span at one spot than were taken that year in the rest of the whole flyway. The volume of birds, alive and dead, has been exceeded only by the volume of birdwatchers who produce Sunday traffic jams around one refuge, by the volume of excited newsprint and letters to legislators, by the anguished cries of farmers scorned and hunters restricted, and by the hours spent by technical people wrestling with what may well be one of the country's most perplexing game problems.

In general we can summarize the continental Canada count this way:

(1) Over the past 20 years Canada goose numbers have increased. (2) The increase seems to be due to the conscious allocation of travel-route and wintering-ground refuges, and to the inadvertent jump in food supply in the wake of mechanical corn-

pickers. (3) But the goose increase is not shared equally by the states in each flyway, significant numbers of birds being "short-stopped" at the Mason-Dixon line. (4) The concentration of birds in certain areas have produced hunter concentrations, with the threat of over-shooting. (5) Because most Canadas breed far above the reach of men and machines, drought and predators, the prospects for maintenance and even increase in numbers is very good. (6) Flock distribution in fall is faulty. (7) The ability and willingness of hunters to abide by selective shooting regulations is uncultivated.

In short, if hunters will cooperate, the Canadas will. The experience of the recent past suggests that we may finally be on the way to successful goose management.

Current Canada Goose Issues

A generation ago, when goose management was in its infancy, we were fairly confident that if we simply sprinkled enough safe havens along the flyways, all would be well. Today we find ourselves attempting to manage geese in a scene of intricate socio-economic factors, of vacuums in basic biological knowledge, and of frequently intense political pressures.

Current issues in Canada goose management can span a rather bewildering array of conflicts. For example, the birds consume an enormous amount of farm grains. Are farmers to be protected from depredations and/or reimbursed for lost crops? Attempting to break up a large concentration of geese at Horicon Marsh a year or so ago, federal personnel proposed to haze the birds south with noisemakers and planes. The agents were publicly threatened with arrest. By whom? By a member of the Wisconsin Conservation Commission. Where does the authority and responsibility of Washington begin and end over migratory waterfowl?

The city of Rochester, Minnesota, harbors a flock of 8,500 giant Canadas from September to March. The Rochester Board of Health has calculated that the amount of fecal material deposited by eight geese in a day is equal to the excrement of one human. When does goose conservation become goose pollution?

Biologists call for a swing away from a few large refuges to the development of many such areas disposed along the flyway. But how do you get the birds to start re-using areas in the south once they have learned to be shortstopped by Yankee corn? There is general agreement that we must reverse the trend toward concentration, depredation, excessive kill, and unequalized recreation. But how do we learn how to manage hunters?

If Canada geese are not to lose their wildness and hence their quality as a trophy bird, contact between human beings and geese should be reduced as far as possible. How can this be made to square with a trend toward opening up refuges to more general public recreational use? The bulk of our goose refuges have been financed by special taxes imposed on hunters. Yet the bird-watching public is growing much faster than the hunting public. How can the costs of goose-management be apportioned more equitably? The goose recognizes no man-made boundaries. To measure his status, control, and mortality requires rather sophisticated cooperative census procedures and harvest quotes among states. How can such arrangements be better conceived and implemented?

These and other problems suggest we are scarcely at the end of the beginning in Canada goose management. While we may have come a long way in our knowledge of Canada geese since 1940, what we still don't know is hurting us as we strive to increase the size of the population and to distribute better the opportunity for bird-watching and hunting.

The Goose Hangs High

We are in a period of significant change in the management of the Canada goose. Our environment is changing. We now have a good basic system of refuges with which to operate. The customers are changing. Birdwatchers are growing in number, hunters relatively in the decline. So our traditions are changing. We seek quality human recreation in many forms as well as geese in quantity for the gun. And our level of confidence is growing.

Where once some might have felt that honkers were on the way to becoming endangered, today we generally recognize that geese offer a great potential for "doing something" with waterfowl.

Our concepts are changing. We believe that Canada geese lend themselves to management by taxonomic, ecological, or geographic units rather than in the mass. Techniques are changing. The modern manager has a kit bag of tested tools with which to measure and manipulate hunters and hunted. In short, we are in an era of transition from general to intensive management of geese and goose harvests. An abiding concern for the resource has become tempered by a growing belief in our ability to maintain, restore, and increase Canada goose populations on a man-dominated continent. Our confidence in turn is tempered by a recognition that we need better research, strong leadership, co-operative and coordinated efforts, and more money.

Larry Jahn, Wildlife Management Institute executive, had this to say about where we ought to be going:

1. We need more and better research, particularly into the special characteristics of distinct subspecies and populations.

2. The Canada goose must be recognized as an international bird that doesn't belong to any one state or province.

3. Something more than gross flyway or even statewide regulations are required; we must apply the "goose management unit" concept:

Where hunting pressure is light, we can count on bag, possession, and season limits to provide sufficient protection; but where hunting pressure is intense, more sophisticated hunter controls will be required to hold harvests within allowable limits and to preserve some semblance of a quality recreational experience. Establishing harvest management units for Canadas will resolve numerous complex problems, but periodic excessive goose numbers and their attendant threats of crop depredation, recreational degradation, and over-shooting will remain a constant challenge to wildlife managers. In short, if we can muster enough facts, funds, and effective teamwork, the future of the Canada goose in North America looks very bright. At least it seems the goose will meet us half way.

Managing Geese and People

One over-riding generalization emerges from the history of the Canada goose: geese and people are a lot alike, and you can't manage one without managing the other.

Like people, geese are flexible, versatile creatures. They will nest, for instance, on trees, banks, sand bars, haystacks, islands, cliffs and in meadows, tundra, marshes, muskeg. Their migration routes are undeterred by desert, mountain, or megalopolis. And their wintering grounds may be a remote swamp or a municipal park. Both species can learn to adapt. When its original Mississippi River sandbar haunts became overrun with brush and hunters, the valley population of Canadas funneled into Horseshoe Lake, Illinois. The hunters followed. Although wild honkers normally have an aversion for tall vegetation of any type, they have learned to feed in standing corn. To help them out, goose managers have learned to plant stunted varieties. At the same time, both people and geese have certain minimum habitat requirements that must be met, if life, liberty, and happiness are to be achieved. For neither species are we absolutely sure in detail what these thresholds of survival are. We do know that both species respond to what may seem like relatively simple habitat manipulations, but neither the direction nor the intensity of change can always be predicted with any certainty.

It is in their moments and sites of stress that geese and people are the most alike. When abnormal numbers of geese and hunters are crowded into comparatively circumscribed areas, strange things take place in both species. The geese tend to lose their traditional caution. What ought to be the proverbial "wild goose chase" can become a shooting gallery situation. Hunters at "inner core" areas likewise fall victim to a mob phychology. Denied the normal rituals of the waterfowler, they concentrate not on more fun per gun but on more birds per trip. The juveniles of both species seem particularly prone to the effects of over-population. It is not too much to speculate that the Horseshoe Lakes and Horicons of the country are actually producing new strains of birds and shooters with unattractive characteristics. Certainly Cecil Williams is right when he suggests that when, in both geese and humans, native traits become too greatly modified by pampering and crowding, something of real value is lost.

Fortunately the Mississippi Flyway story suggests that the cure for the problem of both Canadas and hunters is symbiotic: a system of traffic controls that will regulate the impact of the two species on each other. The activities of geese and people must be more widely dispersed in time, place, and intensity. We probably have the aesthetic sense and the ecological understanding to do this now. What is lacking is the economic commitment and the political savvy. But the improved situation at some sites suggests that we are on the way to enlightened management of both birds and people.

Ode to the Goose Hunter

One thing every sportsman seeks is goose hunting of high quality. It is easier to count geese than it is to measure the quality of recreation. Quality is intangible. Its measurement involves subjective judgements. As William Webb has said, "There is no ruler or scale of absolutes to support our uncertain judgements of what is high quality in recreation." It is often easier to analyze an experience of low quality. In this respect quality is like another abstract concept—justice. We may not be able to define justice, but most of us would agree that we can identify an injustice. Yet each human being does have his own private feelings about quality. Each individual seeks out experiences that he intuitively feels are quality experiences. It is difficult to put into words why a particular experience is of high quality because such a statement involves an attempt to quantify an aesthetic emotional experience. If we are to manage for quality hunting as well as for quantity geese, however, we must make some attempt to describe and delineate as tangibly as possible what it is we seek, as well as what we seek to avoid:

For example, shooting geese in a poorly managed situation approaches slaughter, not sport. There is no suspense, no challenge. The birds are tame, the shooters wild. You are almost as sure of bringing home a gander as if you were to visit your local meatmarket—provided you can run faster than the fellow in the next blind. Goose hunting in such a setting bears more resemblance to a spectator sport like football than to bona fide hunting—a football game between the Green Bay Packers and Gopher Prairie High!

Rightly done, on the other hand, goose hunting is far from a simple, one-time game. It is a year-round ritual, as complete with secret incantations, special garments, secluded chapels, and sacred scriptures as the most elaborate rites of an exclusive fraternity.

The uninitiated hunter may actually regard shoulder-to-shoulder shooting as fun, of course. He may relish the race to claim a cripple, the exchange of cusswords, or even simply his seat in a government-owned stadium. The real question is, should goose managers participate in degrading a great recreational resource in such a manner? The answer surely is "No!" To provide guidelines to quality goose hunting, we have to begin to "tell it like it is."

Capturing the spirit of the sport is not easy. There are many forms of quality goose hunting. There is field shooting—locating the feeding geese one day, digging a pit in the dark of the night, and at dawn the distant gaggles of geese that you try to call in to your decoys. There is the hunt on a sand bar—with the voice of the river in your ears like an organ. And there is the mixed hunt on a duck marsh where geese are more or less an unexpected bonus that makes the morning extra-special. Whatever the setting, certain ingredients are essential.

Wherever he hunts, the principal amulet of the true goose hunter is the decoy, which you fashion in your basement workshop during the dog days of winter. While you may lace your set with some manufactured jobs of styrofoam, you place your real faith in the silhouettes or blocks you rout out by hand, and paint and repaint with all the devotion of a Michelangelo.

An early fall sacrament of the goose hunter is known as "fixing up the blind." No matter how sturdily you may build your marsh hideout in the first place, the amount of annual maintenance is considerable. As you work, you make your distinctive preseason signal, a series of dull thuds accomplished by engaging the butt end of a tamarack post with a heavy mall. Some-

times this basic call is accomplished by a low muttering, not
unlike an oath.

After shoring up the framework with due ceremony, you turn
to the ritual of cutting, bundling, and tying on a new covering
of camouflage. Securing rushes to the snow-fence sides of a
blind is an art known only to waterfowlers. No woman arranging living-room draperies exercises such sophisticated care. From
the perspective of every possible on-coming bird, the outline of
the blind and its inhabitants must be perfectly concealed. To
double-check your artwork, you row out to take a look before
you tie on the last batch of bundles, as if you were arranging
the exact trappings of a communion altar.

At this time of year you will also frequent your marsh shack,
replacing the anchor cords on your decoys or painting your boat
that shade of ineffable beauty, "dead grass tan." Being a true
goose hunter, you are characterized by glassy eyes set in a head
perpetually cocked to left or right, as you listen constantly for
the sibilant sound of waterfowl wings. That is not a cigar you
have in your mouth. It is a reed on which you continuously
practice the cry of the Canada.

The crescendo of the goose hunting ceremony opens at 6:05
on a late fall evening, with the TV weatherman tracing on his
map the course of a big "front" moving down from the north,
preceded by rain and followed by falling temperatures. At 6:10
your phone rings. It is the High Priest of the Hunt—the friend
who has a private marsh. "Did you see the forecast?" he asks.
This is the password. You give the counter-sign. "I sure did."
The High Priest utters the magic words, "Let's leave at 8!"

That is the signal for collecting by the numbers the biggest
stockpile of combat gear this side of Vietnam. Stationwagon
loaded, you flee the suburbs in frantic haste, like a couple of
refugees deserting a doomed city. The rain beats a tattoo on the
windshield, and the wind sends cascades of sodden leaves across
the glistening road. You don't talk much, because idle conversation might break the spell of the ceremony. You turn off the
highway onto a county trunk, then onto a town road, and finally
onto a slippery lane. In a patch of woods you cache the car and
head down through the swamp, your flashlights making only a
feeble dent in the blackness as you slosh along in knee-deep

muck. The rain has stopped now, but the wind keeps up its high-pitched litany in the tamaracks, punctuated once by the gabble of snow geese waving unseen overhead.

After a tortuous half-mile hike you stumble up to a quonset hut. The stubborn padlock finally yields to a special curse, and you enter the mystic domain of the goose hunter—the shack. Lamps are lit, stove stoked, alarm clock set, all according to a routine as immutable as a baptism. You sleep only fitfully, disturbed by visions, and you are up making breakfast before the alarm goes off. You have to crack a film of ice on the water bucket. It will be "a good day for ducks." You paddle out to your blind early, well before the sun is up, because there is a special magic about a pre-dawn pothole. To the accompaniment of the mysterious noises of the night, you place the blocks—a tight knot of puddlers here, a string of divers there, and in between a gaggle of geese. To toss out the decoys you have assumed a kneeling position. This attitude of prayer is a vital part of the goose hunter's ritual. As if in answer to your supplication, a brace of bluebill bursts by.

You hunker down in your blind and watch the dawn wrestle with the dark. A mallard hen becomes audibly enthusiastic back in the rice beds, but you cannot make out what she is talking to. You hear a flock of sprig, pitching pondward, tear the dark silk of the night in one long rending nose-dive, but there is still nothing to see except stars. Then with an unbelievable rush the dark is gone and you are revealed to be suspended in time and space—a pair of goose hunters eye-deep in the marsh like so many muskrats, clinging to the strand of cattails that separates tamarack from tossing waters. A sense of history lies heavy on such a place. Yearly since the ice age it has awakened each fall morning to the clangor of Canada geese and the beating pinions of pintails. In the incredible sweep of millennia that underlies the affairs of nature, your hunt is only a momentary intrusion.

Far to the northeast, winding the oxbows of the river, an arrow of waterfowl cleaves the dawn sky. They appear to be geese. Are they moving steadfastly south or are they looking for a place to sit down? Can they hear you? These are the questions that lend almost unbearable suspense to the world of the waterfowler. You give them a tentative toot on your goose call. They seem to answer. You go to work in earnest. They turn your way, giving tongue to a few querulous honks. You crouch down in the boat and hold your breath. Majestically the flock swings high over you, stepping up its conversation.

Do they like what they see? After an agonizing pass to the west they come pumping back. In a pandemonium of trumpets, croaks, and cries they set their wings and glide down, black landing-gear lowered and rumps white against the far hills. Out in the middle of the pond they come to rest with a great honking and splashing, long necks and beady eyes surveying the scene with the caution that has brought them safely to this spot from Hudson Bay.

This is the climax of the goose hunter's ceremony. No matter how many time you have been initiated, it is always a breathtaking experience. No other outdoor event is so fraught with primeval drama. For one awful second there is nothing in time but you, a little stretch of windswept pothole, and a huddle of wild waterfowl. For a magic moment you look right into the eye of nature. What you see displayed before you, as Leopold wrote, is no mere bird. The Canada goose is the symbol of our untamed past, a cymbal in the orchestra of evolution. His annual pilgrimage is the ticking of the geologic clock. Upon the hunters he visits he bestows a sort of blessing, and upon the place of his alighting he confers a peculiar distinction. Amid the antiseptic surroundings of civilization, a goose pond holds a paleontological patent of nobility. The sadness discernible in some subdivisions arises, perhaps, from their once having harbored geese. Now they stand humbled, cut off forever from the flow of nature.

The geese are not with you long. For a moment they conduct a loud debate over the relative merits of staying or of heading south. The urge to travel wins. With a surge of flapping and gabbling they take off. Their angle of ascent and the direction of the wind are such that they present you with a target as they gain altitude—16 huge honkers spread out in the sky right in front of your blind. The firing of your gun is strictly anti-climactic. Almost reluctantly you throw three quick shots at the milling mass of birds.

One gander collapses with that clean abandon that makes the deed seem preordained. The rest continue on their way without so much as breaking formation, to become once again nothing but a distant "V" on the southern horizon. The geese have fought a good fight, and you have kept the faith.

How your priesthood has developed, anthropologists aren't sure. One theory holds the goose hunter to be a throwback to a race of amphibians that never totally made the transition from water to land. Goose hunters themselves claim they represent the acme in man's evolution, because of their manifest kinship with the sky as well as with the ground. Once the true goose hunter was quite common. Today even in the presence of more geese, he is becoming increasingly rare. Circumscribed by restrictions on his gun, his hours, and his bag; harassed by pseudo sportsmen on "managed" hunting grounds; driven from his marshes by drainage projects, housing developments, and power boats; plagued by the vagaries of the weather; ridiculed by his business associates and ostracized by his family—all in all it could be that the true goose hunter, and not the goose, will go the way of the passenger pigeon, and in another generation will be found only behind glass in a museum. Indeed, maybe the government has been going about things all wrong. Instead of collecting $3 a year from every waterfowler for the protection of waterfowl, perhaps Washington ought to collect $3 from everybody else for the care and preservation of pure goose hunters. Certainly it would be an evil day for America were a certain dawn in autumn not to signal the assemblage of avid goose hunters in isolated swamps, their storied tenacity testifying to something special in the human spirit.

Part III

OUTDOOR RECREATION AND YOU

7. Man, Land, and Leisure

IN CASE YOU HAVEN'T NOTICED LATELY, YOUR LIFE AND ITS LANGUAGE have been changing.

Twenty-five years ago the word "recreation" referred to bowling, card playing, a Memorial Day drive, a Fourth of July band concert in the park, or an ice cream social staged by the local fire department. When a man wanted to chase rabbits on Sunday morning after a busy work week, he went hunting, a sport that at the time had nothing to do with "recreation." Ditto for fishing, camping, and other outdoor pursuits of vigorous nature and rustic charm.

That was 25 years ago. Today, much of that nostalgic pattern of relaxation has been swept aside by a new way of life that has virtually forced itself upon us. Employment has generally been good, and family income has never been higher. We are, as a result, both busier at work and busier at play. We have more money for travel, swimming, hiking, hunting, fishing, camping, and other pursuits that are now dignified by government bureaus under the label of "outdoor recreation."

Ours is a very hurried leisure, chock full of hustle and bustle. Sales of travel trailers and outdoor camping equipment are at record peaks, highways are busier than ever—toting the leisure-seeking family from place to place in search of that elusive "recreation," which is the catchword of the decade. Recreation is still available in individual doses, but it is dispensed most commonly in family-sized packages.

Deer hunting patterns provide a typical example of what has happened. In years past, a man took off for deer camp with his partners for a week or ten days each November. Nowadays,

with express highways ahead and a vacation-conscious family
behind, the deer hunter packs a brace of sandwiches Friday
night, heads north on opening weekend of the season, and is
home ready for work Monday morning. No vacation time is ex-
pended on such a trip. He's saving up for a full bout of outdoor
recreation, served family-style, at the beach next summer, when
the mix of activities will provide something for each member of
the family—fishing, water-skiing, boating, swimming, walking,
and above all else, touring.

In short, large numbers of us have been shifted out of old
patterns of life and into new ways of seeking recreation—changes
that have come steadily in these last years and often without
announcement. Only in looking back over a couple of decades
do the changes show clearly. The sum of the situation centers
around a single theme; namely, that nowhere in this country
of ours do we have enough recreational facilities of the right
kind in the right places to handle the spilling, spreading, growing wants of all our people.

Facing Up to the Land Pinch

Our American bounty of land has limits. In 1900 there were
25 acres of land for each American. By 1950 that was down by
half, to 12.5 acres per capita. By 1970 it was nine acres. The
land pinch around our lakes and rivers is particularly acute and
growing worse. Jams are common. Crowding is the rule. Campgrounds are sardine-packed. The story of playgrounds, parks,
and golf courses is also one of overcrowding. In the last ten
years, for example, the number of golfers has increased by 123
percent, while the number of courses is up only 56 percent. Unless and until we can change these affairs of state, our people
must continue to suffer the wrench of recreational poverty amid
social plenty.

It is not yet too late to solve the riddle, but the day is far
along, and we are behind schedule. Prime lands, particularly in and
around cities, have been taken up for "slurbs" and trash dumps.
Parks become parking lots in cities where urban myopia replaces
urbane wisdom. Patches of trees give way to patchwork con-

struction, over the muffled objections of metropolitan planners, who generally have good ideas but too often are forced to compromise or quit when the coins are down. The only way this steady and continuing and growing trend can be changed is by public clamor and enlightened use of governmental instrumentalities. We can no longer allow ourselves the luxury of doing whatever we please to the land. The only possible course is careful planning of uses made of land, preferably on a regional basis, followed by rational control of land uses based on that planning.

We have traditionally been dedicated to the proposition that every man shall have the right to stand on a piece of ground, to view and enjoy a beautiful scene, to wade a clear river, to gaze in awe at a virgin piece of forest, to hunt a deer, to pitch a camp under the heavens, and to boat upon the waters that we have had in such abundance. We are dedicated to the proposition that every child shall have the opportunity to play on a bit of grass, to cool bare feet in a wading pool or at an open fire hydrant, to enjoy some bit of the out-of-doors, regardless of the place called home. The Lord in all His wisdom, though, only gave us so much water and so much land upon which His people could live, work, and enjoy themselves. There is not room for everyone to own his own piece of open space nor for coming generations to hike and explore unless we share these same waters and lands with each other.

When land control is urged, some local folks argue: "What's the point of all this fancy planning and zoning? Our grandfathers got along without it." Our ancestors didn't face the land pinch.

New Weapons of War

Last winter at Ft. Sill, Oklahoma, some fellow reserve officers and I were exposed to a demonstration of the new weapons of war now in the hands of U.S. Army troops. They make our World War II arsenal seem like popguns. There is the M-16 fully-automatic rifle that weighs only six pounds and fires a .22 caliber bullet at a devastating velocity. The M-79 grenade launcher, an

overgrown shotgun, lobs a small bomb 60 yards. An aluminum
105 howitzer is so light it can be parachute-dropped and man-
handled into position by a couple of GI's. The Cluster Bomb
Unit (CBU) turns a light helicopter into a man-of-war. The
AAARV airborne amphibious vehicle mounts a 257 mm rocket.
The Davy Crockett resembles a stovepipe but gives a front-line
infantry unit an atomic capability. All of these weapons have
this in common: They are designed to render every soldier a
more effective manager of violence, while at the same time exposing him to less close contact with the enemy. They substitute metal for manpower.

Each spring I go to another display of new weapons of war—
the Outdoor Sports Show at the Dane County, Wisconsin, Fair
Grounds. I see no less than 587 exhibits of fishing tackle, boats,
motors, marine accessories, diving equipment, camping gear,
and clothes, including a "camper's village," a tank show, fashion
models, log-rolling, retrievers, diving, archery, trick fly and bait
casting, a fish and game museum, and an acre of trailers. All of
these weapons have this in common: They are designed to render
more efficient our assault on the out-of-doors, while at the same
time exposing us to less and less meaningful contact with it.
They substitute nylon for nature.

Marion Clawson has compared today's outdoor recreation drive
with the "go West" fever of another era. It was 180-plus years
ago that the Continental Congress blew the starting whistle on
the greatest land boom of all time with the Northwest Ordinance
of 1787. In just 100 years the U.S. Census Bureau could no longer
identify a frontier. By any measure, today's rush to the great
outdoors is staggering. If you add up all the reported visits to
parks, forests, refuges, and other recreation areas in 1969, the
total exceeds the population of the United States. We are all part
of the greatest array of peacetime assault troops since the Oklahoma Land Rush.

By common consent, outdoor recreation is "great sport." The
trouble is, it is growing greater and less sportier. A true sport
has a code of conduct. You play it by the rules. You couldn't
employ a super golf ball, for example, even if it were invented,
unless the PGA authorized its use. And you can't tee off without
a permit. But there are no generally accepted codes governing

our invasion of woods and waters. We muscle in, crowd together, pick, slash, and roar in the manner of a heathen horde. We have not yet discovered that, because 20 people can enjoy a beauty spot, it does not follow that 2000 can. It has not yet dawned on us that a true re-creational experience beyond the city limits has practically nothing to do with equipment.

Good Bye, Mineral Point Road

Although it may sound a little bit like King Kong hiring a hairdresser, state and federal highway men are developing an admirable concern for beauty. They are talking particularly about scenic easements. A scenic easement is a device whereby the government pay highway frontage owners not to engage in certain practices deleterious to beauty, or to engage in certain other practices calculated to develop scenery. A common scenic easement is a proscription against billboards. Through its 15-year-old Great River Road project along the Mississippi River, Wisconsin is in the forefront of scenic easement states.

But scenic easements are only a part of an overall beautification program. There are raw banks that can be planted with grass and shrubs, old gravel piles that can be eliminated, old roadbeds that can be camouflaged, streams that needn't be straightened, borrow pits that can be turned into lakelets. Even more substantially, there are park and forest lands that needn't be dissected by highways. There are lake vistas that can be opened up. There are hillsides that can be barred to new construction, or vacuumed of their dilapidated barns.

Highway beautification is an official part of the Great Society, and a good deal of federal money is available for cleaning up and preserving scenery along major roadways, so we are apt to see good progress in this regard. Meanwhile what is happening to our town and country roads is something else again. While we are trying to landscape antiseptic stretches of I-highway, our country roads continue to be robbed of their natural beauty.

A prime case in point is what is happening to Mineral Point Road, a county trunk running due west from Madison, Wisconsin, to Pine Bluff. Yesterday this road made for a pleasant Sunday drive as it followed the contours of Dane County terrain, dipping onto plains, bounding over hills, passing through bucolic settings. Today Mineral Point Road is being turned into a naked speedway, charging through urban sprawl and over the land with all the finesse of a bull. The literal and figurative high point of the drive used to be to top a hill just this side of Pine Bluff and see stretched before you a magnificent panorama of farmland and woodlot, punctuated by the bulk of Blue Mound against the western sky. Now what was once a crest is a cut. The view, and the rollercoaster sensation, are second-rate.

Unquestionably the new Mineral Point Road is a faster way of getting somewhere, assuming you have somewhere to go. It may even be safer. But a bit of natural beauty has disappeared forever. Thanks to the scenic easement device, we have a handle on controlling some uses of private land along super-highways. Maybe it ought to be applied to public rights-of-way along town and county roads as well.

Tourism: Boom or Blight?

How can a tourist region develop its potential as an outdoor playground without destroying its natural attractions in the process? Development resulting from tourism already has blighted many sections of America with garish signs, roadside carnivals, fake souvenir stands, and entertainment gimmicks that fight for the eye and the dollar of the visitor. Such development destroys the very qualities that made the area worth visiting in the first place. At the same time, other areas of the country have used the economic power of the tourist boom to increase and enshrine their natural beauty. Their methods can be applied to communities and rural areas to make them of lasting value both from an aesthetic and an economic point of view.

Rolling green countryside, keen air, lovely dark and deep woods, breathtaking views of open water, seagulls, an aging inn with a friendly tone and an outdoor quality, the rich historical background of the region, water-worn caves, and singing sands— these are the things that make northwestern Wisconsin unique, for example. These are the features that must be preserved and enhanced for the thousands of visitors who will come to see them and stay to appreciate them. The city of Bayfield, Wisconsin, for example, could become nationally known as an authentic Great Lakes fishing community. On the other hand, it could be turned into a cheap tourist trap, indistinguishable from a score of others across the land.

The key to the challenge is local leadership. State and federal instrumentalities can provide technical and financial assistance, and private consultants can provide ideas. But only town and county board members, village presidents, Main Street merchants, and other local people can really determine how a community or a rural area will develop. We need some specific techniques that can be used to guide recreational development—devices like local zoning ordinances, local architectural control commissions, local educational campaigns, and local non-profit organizations that would buy land or supervise land uses to forestall undesirable development. Above all, what is needed is an environmental ethic that will assure the long-term economic and aesthetic well-being of the north, her people, and her guests.

Duck Hunting Is Dying

We sat in a duck blind the other day and watched the death throes of a sport. It was a sad sight.

Duck hunting is the greatest of American shotgun sports. It is demanding of time and talent. It is governed by a complex of rules and rituals. It is fraught with deep anticipation and suspense. Above all, duck hunting is a visual and spiritual experience on a grand scale. The gunner is posted in an immensity of water, sky, and land. Humility seeps into your skin along with the marsh wind. Particularly when you get to your blind early, when the sky is black and the stars close, then it is easy to see that the earth is a star too, adrift in a void. A heavy parka gives little protection against such a thought. It is the vast sweep of sight and sound and sensation, experienced in the utter privacy of a blind, that is the soul of waterfowling. And it is this that is disappearing, not the ducks themselves.

We saw plenty of birds the other day: a single greenwinged

teal hurrying by as if to catch the tailcoat of summer, a knot of ringnecks toying with our decoys, a brace of pintails examining us from great heights, a big blanket of widgeon furling and unfurling over the island before breaking up into scattered flocks rocketing in all directions, a drake canvasback whistling past like a .155 projectile, mallards performing their immemorial minuet over the marsh. But we saw plenty of other things, too: a flotilla of hunters parading to their positions, a boatload of fishermen anchoring just beyond our blocks, hotrodders tooling their powerboats back and forth, a couple of teenagers parking within shotgun range of our blind, cars cruising a highway that was once a town road, a farmer mowing his late alfalfa in what was once a slough, cottagers on their piers where once there were only cattails.

Biologically speaking, any piece of terrain has a fixed carrying capacity. A duck marsh has a very low human carrying capacity. Fill it up with just one too many incompatible people and the sport disappears. That is what is happening to duck hunting. The quality of the shooting is lower than the quantity of the birds. Is there anything we can do about it? Maybe. When the number of basketball fans exceeds the capacity of the university fieldhouse, we ration the seats. We say to each coupon-book holder, "You can go to only half the games each year." In like manner, we are going to have to ration duck hunting. We are going to have to say to hunters, "You can hunt only half the days of the season." And we are going to have to bar our duck marshes to fishermen and hotrodders after October 1st.

Either that, or say goodbye to a great American sport.

A Curtain Comes Down

The curtain has come down on one of the great outdoor shows of my town. The theater was the cornfield that lay to the south and west of Willow Drive at the far edge of the University campus. The time was any sunset. The actors were waterfowl, mostly mallards, tracing out against glowing steppes of heaven the lines of an unforgettable drama. The stage setting was simple yet elegant: To the right, a bay of Lake Mendota, its surface alternately marked by the skittering patterns of windy gusts, or glass-like in its calm; its color horizon-blue, or gold, or green. To the left, a prairie, once wetland, now cornfield, soon to become something else. In the center, a metaled lane following the age-old heave of sand sweeping gracefully between marsh and lake.

Enter the actors, in response to an instinctive cue: the gnawing of an empty craw, or a certain fading light intensity. Up out of the bay the mallards would mount. Singly, in pairs, in clusters they flashed over the willows, swung high above Eagle Heights, stretched out in a long, undulating train, circled, glided, flared, circled again with ancient caution, and finally set their wings to drop in, breasts reflecting the rosy depths of western sky, backs dark against the disappearing sun. That is all there was to the play. The action lasted less than five minutes. Yet while the actors were on stage they sang a wild song that was breathtaking in its beauty and compelling in its strength. There in the midst of metropolitan Madison nature had been able to preserve a momentary hint of America's magnificent outdoor heritage.

It is no more. The lowland cornfield was condemned, and with it the susurrant sound of waterfowl wings. Interestingly enough, the field was not the victim of a super drainage ditch that made for more and better corn. Nor was it filled in to provide sites for ranch houses or laboratories. It was filled up in the name of outdoor recreation, no less. University plans called for a fine playing field, featuring tennis courts, baseball diamonds, running tracks, a pitch-and-putt golf course, and other accouterments of outdoor sports. There was in the equation, however, no place for mallards and birdwatchers. So the dump trucks came and the ducks went.

By all the economic and social standards against which such things are presently measured, the University's new playing field will benefit more people in more ways than could a damp cornfield. There is as yet no cash register or yardstick, you see, to tell us the value of an unobstructed sunset or a mallard on the wing. Someday the scientists may develop such a scale. Whether there will be anything wild, natural, and free left to measure is the question.

Conservation Versus Recreation

How do you increase the quantity of outdoor recreation experiences without decreasing the quality, and how do you pay for it? That is the central question agitating outdoor folks today.

Conservationists are growing more and more edgy over the mass of trampling feet that is subjecting the landscape to a recreational rush. They are concerned lest a growing emphasis on quicky park facilities, cottage tenements, and other tourism developments submerge the conservation gains of the past half-century. They know that outdoor recreation of high quality can be destroyed by the very quantities of people that seek it so desperately. Lands purchased, designed, and dedicated to hunting and fishing can be preempted for mass-use parks. Lands set aside as wilderness can become neon-lighted tourist traps. Acres in county forests can be diverted to cabin sites. Long-established policies and principles can give way to hurry-up needs. Complicating the whole question is the fact that no one in the conservation business has any real standards by which to measure the "quality of outdoor experience." For an oldtimer, it may be a wild, unpeopled swamp, and straw-tick bedding in a ramshackle deer camp. Youngsters, on the other hand, may be satisfied with the out-of-doors as it is today, having never known an old deer camp nor a stretch of untouched trout water. Perhaps quality outdoor experience is what you get accustomed to at an early age. One thing we do know for sure is that for many people high-density outdoor recreation has the capacity to destroy the very enjoyment it sets out to capture. The number of fans at a football stadium does not detract from and may even enhance the sport. The same cannot be said for grouse hunting on a back forty or even for picnicking in a park.

The solution to the dilemma is but dimly seen. Throughout the nation no conservation leader has yet been brave enough to offer any hard and fast answers. At a recent national meeting some proposed solutions included zoning to limit certain uses to certain areas, heavier taxation to support more and better widespread outdoor facilities, licenses that restrict use of wilderness areas, ordinances against abusive practices, education programs that teach people how to appreciate nature without consuming it, and shifts in vacation scheduling to allow better use of spring and fall months. All these proposals say one thing in common: "The old ways of free and unrestricted use of the outdoors are going. Planning, taxation, and restrictions must pull in double harness with growth of outdoor activities." Will we get combined federal, state, and local action strong enough to meet the challenge soon enough? The answer probably is "yes" in some areas, "no" in others, and a qualified "maybe" in much of the country. The result to the outdoor fan may be a patchwork of pleasure and frustration.

The Gregarious Hermits

Who are the people who flock throughout the country as vacationers in summer? Who are these creatures who seemingly seek surcease from the city only to crowd into campgrounds or cruise the super-highways? A recent study by Professor Isadore Fine of The University of Wisconsin School of Business sheds some interesting light on the genus tourist.

Half of them in Wisconsin are Wisconsin natives exploring their own state, and half of them are non-residents. Half of them come from urban areas, half of them come from the country and small towns. Only a third of them are professional people, over a third are unskilled workers. The average age is 35 to 44 years. As many have not graduated from high school as have graduated from college. As many have an income of less than $3,000 as have an income over $20,000. The one-day trip is the great common denominator of vacationing. Engaging in water activities, visiting natural sites, and picnicking were the common activities on these excursions.

Do people really want to get away from it all on their vacations? The majority told Professor Fine that they did, but they also said they wanted to have all the comforts of home. "It is hard to reconcile such statements into any meaningful conclusions," he says. "Perhaps this is a phenomenon of big-city living under which many individuals may very well believe that a few acres of woodland constitutes wilderness."

What do vacationers like about Wisconsin? Scenery and sight-seeing, parks and waysides, fishing, and good roads. What do vacationers not like about Wisconsin? Lack of parks and way-

sides, poor camping facilities, expensive resorts, poor roads,
poor fishing.

It's hard to draw any meaningful conclusions from these responses, either. Ditto for the following: Queried for recommendations on park improvements, some tourists asked for better rest
rooms and others asked for a more natural environment. There
was little criticism of the fees charged for the use of the state's
parks, incidentally. Some respondents volunteered that they
would pay higher fees.

Thirty percent of the respondents indicated they visited relatives on their trips, but only 20 percent said that relatives visited
them. "One must leave it to the psychologists to explain this
divergence in responses," Professor Fine comments.

All told, what emerges from this study is a picture of an
ordinary slice of Americana on wheels.

Research Plays Funny Tricks

In some ways, the world was a happier place before the age
of research. We could nurse our illusions with impunity.

Take, for example, the long-held belief that if you could just
bring enough people up to an acceptable level of knowledge
about biological and social facts, the wise conservation of natural resources would automatically follow. Now along came Professor John Ross and Keith Stamm of the University of Wisconsin Department of Agricultural Journalism with a study that
indicates "opinion on any given conservation issue will be shaped
by factors other than or in addition to a general sophistication
in matters of conservation."

The detergent problem, thought happily disposed of when
manufacturers switched to the "soft" variety in 1965, may still
be a problem. Recent research at the University of Georgia indicates that the new detergents do indeed break down under bac-
terial action far more readily than did their "hard" predecessors.
But the new variety still can be very troublesome when sewage
treatment is inadequate, as in most septic tanks.

Up in Montana, the Bonneville Power Administration is using

silver iodide generators in an effort to augment snowpack in a 2,000-square-mile drainage basin upstream from the Hungry Horse Reservoir. If they can pull snow out of passing clouds, the resulting water would be a boon to northwestern hydro-power production. And if they succeed, the Montana Department of Fish and Game contends, the work will strike a damaging blow to elk and other big game in the Hungry Horse basin. Deeper snow could force the animals into yards where they would over-browse available natural forage.

Since our days of reading Thornton W. Burgess, we have liked to think of wildlife occupying pocket-size pieces of home range like humans do. This anthropocentric fable has been blasted by Gerry Lynch of The University of Wisconsin Department of Wildlife Ecology. Studying raccoons and their habits on a Canada duck marsh, he fitted a number of animals with light-weight radio transmitters. One animal sent signals for a day and then disappeared. He was later trapped by an Indian—no less than 175 miles away.

A couple of years ago we were told that if elm trees were injected with the chemical TCPA, it would strengthen their resistance to Dutch elm disease. A University of Wisconsin plant disease specialist now says that although TCPA was a promising treatment in nursery trials, effectiveness of broad-scale treatments is unpredictable. We still don't have a chemical that will effectively control Dutch elm disease without contaminating the environment.

A generation ago we read that biologist, metallurgist, and ballistic experts were pooling their knowledge in an effort to reduce or eliminate lead poisoning as a mortality factor in waterfowl. The solution continues to evade us. Waterfowl pick up lead shot from the bottom of marshes. It lodges in their gizzards where it creates toxins. Now a new joint $100,000 investigation is searching for a substitute for lead shot.

Ten years ago we were all sure trout were goners in the Great Lakes, the victims of sea lamprey depredations. Now the eel-like lampreys are under control, thanks to research, and the lake trout are bouncing back dramatically, only to become contaminated with DDT.

What Does the Public Want?

Determining and satisfying all the divergent demands encompassed by outdoor recreation will be the neatest trick of the week. Outdoor recreation is not a nice, neat, homogeneous product; it is composed of a wide variety of essentially competitive activities and values. To protect in some way the rights of each outdoor recreationist, we can't apply the classic conservation slogan, "the greatest good for the greatest number." This leads us into the trap of using unsophisticated public opinion polls in which we ask the pollee to check, in order of his preference, a list of the common kinds of recreation that happen to be easily definable and available. We get such inane answers as "driving for pleasure" because nowhere on the list are such items as "sitting on a log surrounded by silence."

What happens when you do a careful job of surveying the wants and needs of outdoor recreationists is illustrated by the study recently made of lakeside residents in southeastern Wisconsin. It turns out that only one out of five cottagers in the Fox River watershed actually goes fishing. This includes both vacationers and year 'round residents. But this doesn't mean they don't place a high value on their lakeside setting. As one man put it:

"My major outdoor activity is just looking at the lake."

And what were the problems of these people? Almost 25 percent of those contacted cited improper boating—too many water skiers and too fast power boats—as the main problem. All of which suggests that in the heart of many outdoor recreationists there is an aesthetic factor controlling enjoyment. In the final analysis, high-density outdoor recreation has the capacity to destroy for many the very satisfactions they set out to capture. The price of overuse can literally be the physical deterioration of a lake, as when too many crude septic tanks befoul the waters. But even more destructive can be the psychological degradation that sets in when the human spirit is assaulted by a pall of smog, the blare of transistor radios, and the snarl of outboard engines.

The Abominable Snowmobile

If you have ever wondered what it was really like when Henry Ford's Model T revolutionized America back when grandpa was a boy, wonder no more. Just go north before the winter is over. The north ain't what it used to be. The snowmobile is here, and it has turned the ways of man and beast upside down. You simply have to see it to believe it. I spent a couple of days in Forest County, Wisconsin, recently and I haven't recovered yet.

I suppose it is the sheer numbers of the machines that stagger you initially. Down in southern Wisconsin, a snowmobile is still something of a novelty, but not up north. Everybody, but everybody, it seems, has a snowmobile, whether you are the town doctor or on relief. In fact, you don't really have any status in Crandon or Armstrong Creek unless your family has two snowmobiles. They are to be seen everywhere—not just on prescribed trails, but everywhere: swishing over drift-topped meadows, along frozen waterways, on county trunks, and even on city streets. And these machines are no longer simply mildly motorized sleds. They come in big, bigger, and biggest sizes, capable of speeds up to 80 miles an hour, and sporting such exotic names as Charger, Diable Rouge, and Panther. They have streamlined hulls, headlights, self-starters, and padded rumble seats.

The snowmobile is not simply a means of transportation in northern Wisconsin. It is a way of life. Snowcruisers use their sleds for explorations, family outings, hunting, fishing, rallies, and derbies. They are workhorses for farmers, loggers, and wardens. For modern teenagers they are the buggy with a fringe on top. For young adults they get you around to all the night spots in a hurry with the gang. For oldsters they are the difference between sitting by the fire and getting outside. And of course you can't be seen on a snowmobile in ordinary clothing. You have to have special coveralls, boots, helmet, and goggles. The sights of snowmobilers are overpowering enough, but it is the sound and smell that really get you. One afternoon on the shores of Lake Lucerne I tried to find a five-minute period that wasn't interrupted by the roar of a motorized ice fisherman, but I had to give up. The smoke from multiple exhausts hung like a blue haze over the ice.

Ernie Swift, former director of the old Wisconsin Conservation Department, saw it coming. Just before his death he wrote: "Now comes the snowmobile to sputter and snort through the winter solitude and destroy the last vestige of isolation. Little of the out-of-doors can be appreciated from the back of one of these gadgets. All wildlife flees before their fuss and fumes, and the very essence of nature is diluted."

Thousands of snowmobile enthusiasts obviously disagree. For this growing clan, the surging machines add an exhilarating new dimension to winter. They are content to see nature flit by fast. But for some of us something of value has been drowned in the sound of their motors.

Carrying On the Traditions

In the decade ahead, resource development personnel will face pressures that make the conservation conflicts of old seem like cream puffs.

At the outset there is the sheer pressure of population. There is the pressure of industrial expansion. There is the pressure of agriculture, attempting constantly to put new lands under the plow. There is the vacationland pressure. And there is the steady pressure of sportsmen, who continue to expect something to catch or shoot at, and a place to do it.

If these mounting pressures were non-competing, the job of planners would be fairly clean-cut. But modern land-use alternatives are rarely simple black-and-white propositions.

The manifold factors that must be taken into consideration in state resource planning were brought strikingly into focus with the publication of a report by a special Wisconsin study group on the Wolf River basin. The Wolf rises in Pine Lake in Forest County some 25 miles south of the Michigan line and drains about 3,750 square miles of forests and farms on its 223 miles south to the Fox River ten miles upstream from Lake Winnebago. It is Wisconsin's least harnessed major river. What should be its future? Should it be developed for hydro-electric power? The dams would block famous pike and sturgeon runs. Should big flood-control dams be built? The cost wouldn't necessarily be worth the damage alleviated. Should Wolf water be taken for potato irrigation? Power plants and trout fishermen object violently. Should water be impounded to stabilize summer flow? If so, do you cater to the canoeists in the white water of the Menominee Reservation or to the power-boaters in Lake Butte de Morts? Should industry be encouraged in river cities? Pollution could ruin riparian values. Should the Wolf be left in its relatively natural state? Who then pays the bill for more parks and public access?

And you can't stall around in facing these issues. In the absence of an overall plan, local and state agencies play it by ear. Whatever plans you come up with, how do you get 10 counties, 91 towns, 29 villages, 8 cities, 6 federal agencies and 13 state bureaus to agree on the future of the Wolf? Multiply the Wolf basin by scores of similar watersheds in the state and you get an idea of the problems faced by the Natural Resources Board. Sportsmen have one key hope: that planning personnel will carry on the great conservation tradition—"to provide an adequate and favorable system for the protection, development, and use of the forests, fish and game, lakes, streams, plant life flowers, and other outdoor resources."

Substitute for Nature

As opportunities for old-time outdoor recreation decline, outdoor fans introduce substitutes. Sometimes we substitute targets, sometimes sports themselves.

One of the first substitute targets to be introduced was the brown trout. Even before the turn of the century, it became clear that our native brookie was no match for what was happening to our streams. All sorts of man-made intrusions caused gravel spawning beds to disappear and water temperatures to rise. So early fish managers introduced the brown trout from Germany and Scotland. Not being so fussy about his habitat, the brown trout has thrived. Today we seldom even think of him as an immigrant, so Americanized is he. Another very successful second-stringer is the ringneck pheasant. When native prairie chicken and quail populations virtually disappeared, we stocked the ringneck. He filled what the scientists call an "ecological niche." Today the pheasant is a principal upland game bird, and his hunters are legion.

With the increasing disappearance of rural open spaces in which to roam, we are presently seeing on a mass scale the substitution of the sport of golf for old-time outdoor diversions. Courses are being developed on the outskirts of our cities for the growing numbers of yesterday's fishermen who today are wielding nine-irons instead of fly rods. For the veteran outdoorsman, golf has certain limitations, to be sure. The landscape is well manicured, not wild. You have no bird or fish to blame if you do not score well, only you yourself. You habitually play with a foursome, so you cannot strike out alone and return with a tall story. But golf at least gets you outdoors easily and regularly, and it allows you to identify yourself with those new folk heroes of America, Jack Nicklaus and Arnold Palmer.

What is more, golf has some soul-satisfying rewards of its own. It is deceptively simple and endlessly complicated. It requires complete concentration and total relaxation. Especially in the spring of the year, when the first warm sun presses down on your shoulders, when the grass has just been mowed for the first time and sits there damp and green, with its fresh-cut newly fertilized smell filling the air, when the sky is a deep blue roof punctuated only by an occasional puff of cotton, a golf course is as intoxicating a place as a stretch of trout water. What the hunter finds in the flight of a mallard the golfer finds in the flight of a good drive—the white ball sailing up and into that blue sky, growing smaller and smaller, almost taking off in orbit, then suddenly reaching its apex, curving, falling, describing the perfect parabola of a good hit, and finally dripping to the turf to roll some more.

Unreconstructed outdoorsmen still look down on golfing as a poor substitute for less formal, more rugged outdoor recreation. But you can say this at least for golfers—they pretty much pay their own way. Maybe when bird-watchers and bass fishermen start paying a dollar or so for every round of their sports, we'll begin to make a dent in the acquisition of adequate open space for them.

Sideroads to Nowhere

A couple of years ago I wrote a book called *Wisconsin Side-roads to Somewhere*. Its theme was an echo of Zona Gale, ex-tolling the virtues of "a serene sideroad existence." You can't hear or see the important messages of the out-of-doors if you stick to the superhighways, I wrote. You have to sacrifice speed for scenery and wander around on sideroads. One of the side-roads I particularly recommended was Dane County Trunk BB, running east from Madison—skirting Blooming Grove, Cottage Grove, Deerfield, and London—to Lake Mills.

Thirty years ago, commuting weekly between Lake Mills and Madison, I got to know this road intimately. It was a friendly road, you know what I mean? It twisted and turned through drumlins and swales, every curve and every rise revealing a pleasant farm vista, a surprising expanse of marsh meadow, or the filigree of oak and hickory against the sky. There were no manicured shoulders. Sumac and dogwood and plum and le-gumes pressed in on every side. You had to go slow, and you wanted to. At least once a year I like to perform a pilgrimage along BB, even though I can get where I'm going ever so much faster on an I-road. I chose a recent Sunday for such a drive—and I got the shock of my life. County Trunk BB has been "improved."

I suppose through an engineer's eyes the road is now a work of art. For me it is ruined. There are no more dips and crests. The right-of-way is now as flat as a beltline. There are no more intriguing twists, with their rewarding views. The right-of-way is now as straight as an avenue. There is no more roadside vegetation. The barren shoulders are 50 yards wide. Farm front yards, country cemeteries, rock formations, potholes, groves, church lawns—all have been bulldozed away by the dull, gray concensus of "progress."

Highway engineers will doubtless say that County Trunk BB is now much safer. I dispute that. The old, twisting road forced you to limit your pace. The new one is an open invitation to step on the gas. We have turned a dramatic sideroad into a dangerous dragstrip. There might be some point to re-tailoring

County Trunk BB if it were really necessary in order to provide measurable numbers of people with adequate access. But it wasn't necessary at all. County Trunk BB parallels a 4-lane highway. Scarcely a mile separates the two routes. If you wanted to make time, and everybody does on occasion, you could always take the throughway. If you wanted to meander, you could take the sideroad. But no more. BB is the victim of segregated highway funds that seemingly have to be spent without regard to anything but the profits of the bulldozers and blacktoppers.

Somehow or other we've got to blow the whistle on the highwaymen. Somehow or other we've got to convince them that a straight line is not the shortest distance to the quality of a drive in the country.

GND Versus GNP

It was 30 years ago that a state highway department engineer drove up to our farm in Jefferson County, Wisconsin, to inform us that a new road was about to be laid between Madison and Milwaukee, and that our lower 14 acres of tamaracks, pitcher plants, and marsh wrens stood in the way of progress. Today, as you tool along I-94 near Lake Mills, you traverse what was once the Schoenfeld arboretum, now a four-lane cement slash. Along the bare shoulders of the road you are apt to see a new generation of highway workers planting trees and shrubs, spending some of the million dollars in "highway beautification" funds the federal government allocates to Wisconsin annually to restore what could have been saved in the first place.

Interestingly enough, both the cost of a highway project and the cost of an effort to restore the scarred landscape get added into our "gross national product," that yardstick of overall economic activity. This suggests there are many elements of the human condition that GNP cannot measure, "gross national deterioration," for example.

To try to correct the situation, various government bureaus are trying to come to grips quantitatively with what the American Association for the Advancement of Science at its recent meeting called "the threatened collapse of our environment":

"The earth's supply of available oxygen is being depleted.
. . . Large bodies of water are steadily being fouled. . . . In innumerable other ways the precarious balance of nature is being
disturbed."

In the Department of Health, Education, and Welfare there
is a new Deputy Assistant Secretary for Social Indicators whose
staff is trying to report regularly on the quality of life in the
United States. In the Department of Interior, the Secretary now
annually totals up the numbers of acres of land acquired for
public use in forest, park, open space, fish and game, and multi-
purpose reservoir areas.

But the unplanned diversion of lands to urban and highway
development goes on. It was nice coincidence that President
Johnson's Conference on Natural Beauty took place at the same
time that the forces of progress were bulldozing through historic Morristown, N.J. The good ladies of Morristown staged a
sit-in demonstration to declare that their town's beauty was
threatened not by billboards but by the highway itself. "These
women," a state highway department spokesman said, "just
aren't looking at the big picture."

They probably weren't—perhaps because the big picture
threatens to show a land striped with concrete scars, a land that
permits no trees except the saplings of the official landscaper,
no houses except the condominiums of urban renewal, no beaches
except those with lifeguards and litter baskets. It is a picture of
"natural beauty" in which all that is natural—the old, the untended, the surprising, the vine-choked, the animal-tracked—has
been painted out by the brush of GND.

Of Mice, Men, and Mayhem

That our country has simply been unable to solve many problems associated with urban life, recent inner-core riots are abundant evidence. Not the least of the problems is the lack of outdoor recreation opportunities for the slum-dwellers bypassed by
the affluent society of which they are merely onlookers. For these
people, idleness replaces leisure. It is not that they are discrim-
inated against in parks and resorts. The problem is that chronic

poverty prevents thousands from escaping the city to enter the mainstream of American summer life.

We are two societies living side by side, rather than together. All the resources of our adaptable, inventive, generous, rich land have been unable to weld this tragic schism. With luck, some people of the ghettos may find some periodic recreation on the patches of concrete called playgrounds, which substitute for parks in megalopolis. But for most of these people, recreation on even the simplest levels is now an aching desire. Indeed, decent open spaces for play may be much more than something which would be "nice" to have. There may be an ecological limit to the carrying capacity of a piece of urban real estate. Studies of the effects of crowding on animal populations at least suggest that tenements may have within themselves the seeds of inexorable convulsion.

Over-crowding causes some animals like the lemming and the gray squirrel to go on heedless migrations. When the clapper rail becomes too crowded, he invades a neighboring rookery to destroy eggs. Over-abundance in woodchucks seems to be accompanied by a reduction in the numbers of eggs ovulated by the female, and even by an increase in numbers of embryos resorbed. Population stress in snowshoe hares results in tremors. Deer crowded in winter yards browse out the sheltering swamps. When concentrations of lake plankton reach a certain level, the minute plants give off chemical substances that inhibit or kill their neighbors. When goldfish become too crowded, they will not spawn. Other overcrowded fish species eat their own kind. The dramatic fluctuations of cottontail rabbits and ruffed grouse are attributed by some scientists to physiological effects from crowding. Recent controlled experiments with the crowding of mice have revealed that these creatures undergo a form of excruciating stress under such conditions, resulting in impotency, homosexuality, or even death.

Who can say that the violence that has rocked America's cities each summer is not the inherent response of the human animal to unconscious mental and physical crowding in the absence of enough "get-away" opportunities?

Cultures in Conflict

Down at the southern tip of Lake Michigan they're wrestling with one of the toughest problems in the history of the National Park Service: how to protect a vignette of primitive America and yet provide pleasure grounds for more and more people, all in a unique natural area that lies square in the path of a monstrous urban avalanche.

The new Indiana Dunes National Lakeshore encompasses some of America's most precious geological, ecological, and cultural artifacts. At the same time it borders black ghettoes, WASP suburbs, steel mills, jetports, I-highways, glossy shopping centers, rich farmlands, and rural slums. On the ridges of the Dunes is to be read in bold relief the story of the glaciers that carved out the Great Lakes. In the bayous of the Dunes are to be found the living museums that led Professor Cowles to define the science of plant ecology. In the log houses and field burying grounds are the evidences of the great westward trek by voyageurs and immigrants. Nearby are hundreds of thousands of people of all colors and races yearning for a Sunday at the beach, and great furnaces yearning for the ore that comes down from the Mesabi.

In 1916, when Stephen Mather first proposed a Dunes National Park, there was a decent chance to lay out a well-balanced region at the toe of Lake Michigan. Today the Calumet watershed speaks of nothing so loudly as the tragedy of heedless human sprawl. Yet Congress has decreed that the National Park Service must make of the Dunes a viable conservation-recreation area, and the Park Service will try. How to protect remaining natural amenities without penalizing appropriate use, and how to develop reasonable recreational facilities without destroying the dunes themselves—these are the vexing problems that make of Dunesland the epitome of America's land-use dilemma today.

Parklands Versus Taxes

"They're taking our land off the tax rolls!" is the frequent complaint when state or federal bureaus move into rural areas and acquire property for public recreational purposes. What in fact does happen when a state develops a new park, for example?

Do local residents take a tax beating, or do they reap financial benefits?

What has happened in Dodgeville (Wisconsin) Township as a result of the creation of Governor Dodge State Park is revealed in a recent survey. The state began acquiring lands around Cox Hollow in 1950, and by 1964 had pieced together about 4,000 acres with a total assessed valuation of $128,101. Between those years the local tax rate, in Dodgeville Township rose by 97 cents per thousand dollars of valuation. So on the face of it you might say that every resident of the township was paying a dollar more a year for every thousand dollars worth of property to make up for the loss of Governor Dodge from the tax rolls.

But it isn't quite as simple as that. Over half of Dodgeville Township's budget goes for school expenses, and during the period in question a new senior high school was being funded. This was the major cause for increased taxes, not the loss of taxable land. Meanwhile other things were happening. Near the park entrance sprang up a root beer stand, a park manager's residence, a cheese mart, a restaurant, a motel, and two homes. Other developments are on the drawing boards. It seems reasonable to expect that such improvements will add enough increased valuation to the township tax rolls that by 1970 the value of all new developments will exceed the value of all lands lost to the park. Over 200,000 people visited the park in 1964. It is estimated they spent between $430,000 and $575,000 in the vicinity of Dodgeville. This amount could double by 1974. Another benefit: the Soil Conservation Service estimates that flood control benefits to the township from park water control structures amount to $1,900 annually.

The results of the Neuroth study seem pretty clear: The long-term financial effects of a park development are all to the good of the local economy. The short-term effects on a particular township can be measurable. The people who pay are not necessarily the people who get the benefits.

Gradually the state and federal governments will likely work out a sliding scale of reimbursements to local areas that will ease the initial shock when lands disappear from the tax rolls but that will not continue when the benefits from public property begin to exceed any loss.

8. Biopolitics

TIME WAS WHEN A CONSERVATIONIST COULD BE REASONABLY EFFEC-
tive as a lone operator. If he were a hunter he could ration his
take of game. If he were a fisherman he could put back the un-
derling trout. If he were a farmer, he could strip-crop his slopes.
If she were a housewife, she could eschew bird feathers on her
hat. There is still a place for personalized conservation, of course,
but the real decisions in conservation today involve collective
action, so pervasive and so monumental are the assaults on the
quality of our environment. Which means that conservation has
moved more and more out of the arena of the home and the
town meeting and onto the stage of national decision-making.
So that when somebody asks, "What can I do for conservation
today?", the answer has to be, "Vote for a Congressman or a
Senator with a conservation conscience!"

The need for such action is clear.

Each of us can recall a favorite picnic spot that is now a
housing development, a wonderful duck marsh that has been
drained for farming, a scenic highway stretch that is now clut-
tered with billboards and roadside stands, a once-secret trout
stream that is no more, or a peaceful lake that has been encircled
and closed off by private landowners, or overrun and heavily
pressured by powerboats. Our failure to act will do far more
damage to America's future than that inflicted by the lumber
barons when they thoughtlessly destroyed our virgin stands of
white pine and hardwood 75 years ago.

The "harsh facts" are written large on the landscape of a state
like Wisconsin: Camping has increased 246 percent in the state
parks in the past ten years. Carp now occupy lakes that once be-

longed to game fish. Most of the trout waters in Wisconsin are destroyed. Filling, dredging, and pollution threaten to destroy 50 percent of the northern pike spawning grounds in the next decade. More than half of our original five million acres of wetlands have been drained in the past 60 years.

The public is ready and willing to put up hard cash for environmental quality. Such evidence comes from a recent national Gallup poll conducted for the National Wildlife Federation. Some 85 percent of the respondents said they were "concerned" about environmental degradation. When asked to put their money where their mouths are, three out of four said they would be willing to pay extra taxes for conservation. Three out of four said more land should be set aside for conservation purposes. No less than six persons in ten said they would be willing to serve on conservation committees if they were asked to do so.

Never before, says Director Tom Kimball of the Federation, has there been less justification for a wringing of hands over public apathy. Never before has there been less justification for congressional or legislative footdragging. Now is the time for conservation leaders to put up the officers to lead the charge against environmental deterioration. Now is the time for politicians to read the hand-writing on the wall.

At the same time, in a day of massive federal and state programs for outdoor resource conservation, redevelopment, and maintenance, it is easy to begin to think that there is no longer any place for the individual citizen to exercise any meaningful responsibilities toward the enlightened custody of the natural environment. Such, however, is not the case. There is, if anything, an even more crucial role to be played by the man-on-the-street or the man-on-the-land if gross governmental goals are to be translated into action on the landscape.

State Associations of Soil and Water Conservation Districts, for example, annually pay tribute to individual farm owners or operators who do an outstanding job in fish and wildlife conservation. Recent winners testify to what can be done by individual farmers with a sense of husbandry. One of them is a neighbor of mine, Roy Boberschmidt. An Oscar Mayer executive, Roy purchased an Iowa County farm four years ago. Since then he has done an outstanding job of habitat improvement. He has

fenced, sloped, and seeded over 7,000 feet of Trout Creek. He
has constructed three cattle crossings in the creek with gravel
for trout spawning grounds. He has dug two trout ponds and has
a third under way. Roy has converted all his cropland to grass-
land. He maintains his fence rows in brush and shrubs, and he
is planning to plant 4,000 white pine and spruce trees. His 2,600
feet of ditch banks he maintains in grass for wildlife nesting.
He protects 217 acres of woodland from grazing, as well as 30
acres of marsh. All of Roy's land is open to the public for hunting
and fishing.

But you don't have to own or operate a farm in order to practice sound conservation practices. In an increasingly urban society, most people find their conservation opportunities at the
polls or in more direct political action. When the Outdoor Resources Action Program (ORAP) extension bill was up for public
hearing in my state capitol recently, for example, the room was
filled with housewives and businessmen taking off from their
busy days to testify in favor of the acquisition and development
of more public lands for public management. And no longer are
the voices of conservation in Wisconsin limited to conventional
sportsmen's organizations. Conservation programs have a whole
new list of publics. Typifying the "new wave" in conservation in
Wisconsin was a Green Bay housewife representing the Mayor's
Committee for a Cleaner, More Beautiful Green Bay.

It remained, however, for another form of individual action to
set the most significant tone at the ORAP hearing. Madison's
Assemblyman Norman Anderson, speaking with his usual force
and acumen, pointed out that the real purpose of ORAP is not
simply to increase recreational opportunities. "When we degrade
our environment, we demean the human spirit," he said. "Our
goal is not just to halt the destruction of natural resources, it is
to uplift people."

The Second Battle of Antietam

Some bass streams are haunted as surely as a cemetery. Their
water tells a hallowed story as it quavers over sun-dappled rocks
or flows in deep green mystery beneath a shelving bank. The

trees remember. When the wind blows, their leaves rustle as the echoes of never-quite-forgotten shouts. It is so with Antietam Creek. This beautiful Maryland stream is a prime maneuver area for modern bass fishermen and sightseers. Yet something of the past lingers on, in a rusty CSA buckle picked up here, or a Minie ball buried in some old bur oak. You almost expect to meet a platoon of Pleasanton's cavalrymen watering at your stretch of the river. To fish the Antietam is to fish with ghosts.

Just how many ghosts haunt these Old Line hills is shocking, particularly compared to casualty reports from Saigon. In just one afternoon of pitched fighting along the Antietam around Sharpsburg, nearly 25,000 boys were killed or wounded. It was in the second year of the Civil War, following his smashing victory at Second Manassas, that General Lee decided to invade the North. On September 17, 1862, General McClellan came up over South Mountain to strike Lee at Sharpsburg, with Wisconsin's celebrated Iron Brigade in the vanguard. Four successive Union assaults forced the Gray lines back to the Potomac but did not carry the field. Lee left his campfires burning that night and slipped away across the river. The battle, and the National Monument since laid out there, get their names from Antietam Creek, which rises near Hagarstown and flows under Burnside's Bridge, where a Northern general by that name made a halting attack, on to the Potomac. You can fish for the smallmouth bass that abound in the Antietam, or you can spend the afternoon looking at all the statues and plaques along Bloody Lane and on John Brown's farm. For this is hallowed ground.

It is hallowed, that is, to everybody except Potomac Edison. The power company has laid out a right-of-way that cuts straight through the area and is proposing to desecrate these historic hills and valleys with 110-foot-tall towers. The proposed power line has been condemned by everybody from members of local zoning boards on up to Interior Secretary Stewart Udall. But nobody apparently has the power to stop the power company. It is all strangely reminiscent of Wisconsin's recent ill-fated fight with Northern States Power, which is erecting a plant on the St. Croix River against the pleas of local, state, and national figures who wanted to preserve that river from the onslaughts of coal barges and heated water. There can be argument with the country's multiplying needs for electric power; the demand doubles every ten years. But it is a perversion of justice to allow power companies to be the final judge of what values shall be lost to the march of the dynamoes. A number of bills now before Congress would help correct the situation.

We lost the battle of the St. Croix. We should not lose the second battle of Antietam.

Vote "X" for "E"

Voters of a growing number of cities have a chance to "do something" about the sad state of the environment. It isn't much, in a way. It may be more ceremonial than significant. And yet it's a start. What you can do is vote "yes" on the so-called "environmental" referenda that are showing up on ballots. A typical referendum says:

> Shall it be the policy of the people of the city that we have a right to a clean and healthy environment; which right has priority over any use of the environment for public or private ends; that the city demand and achieve an end to the degradation of the environment through all powers available to it and through the advocacy of improved environmental control programs at the county, state, and federal levels of government.

Now that is either a very innocuous or a very dramatic statement, depending on your point of view. It doesn't call for any shotgun action, but on the other hand it makes an important point in an unmistakable way. In other words, it's sort of like the Declaration of Independence. A leading lawyer has made the following analysis of such a referendum:

"The environmental referendum provides a mandate from the electorate to establish a basic policy framework within which the city council should consider issue that will affect environmental quality. Its effect is one of persuasion rather than legal compulsion, since the referendum is advisory, not mandatory, and is directed toward general issues. The fact that it is not legally binding on the city council should not, however, diminish its importance as an expression of the desires of the citizens in instructing the council to take steps to improve environmental quality."

So if you are one of those who agitated about water pollution, air pollution, shrinking open space, noise, and blight, at least go to the polls when you can, and put your "X" on the line.

The ASCS Goes to Town

Something called the ASCS is coming to the rescue of my Lake Mendota. To city folk the ASCS may seem like a strange new alphabet animal. But to farm folk ASCS is an old friend. ASCS stands for Agricultural Stabilization and Conservation Service. It is a federal program with headquarters in Washington and offices in every county. Its ancestor is the AAA of Henry Wallace. A primary mission of the ASCS is to pay farmers for employing certain agricultural practices that are declared to be in the long-term public interest. For example, if I were to fence a hillside woodlot to keep the cows out and thus to cut down on erosion, the ASCS would reimburse me.

Sad to say, not all ASCS-approved practices have been in the real interests of conservation. For instance, most of the wetland drainage that has decimated Dakotas' "duck factories" has been subsidized by the ASCS in the name of improved grain production.

At any rate, the ASCS is an existing piece of social machinery, any now it is beginning to tackle the problem of agricultural pollution. My Lake Mendota is getting increasingly polluted. Mainly, it gets too heavy a load of nutrients and sediments, leading to a variety of unpleasant and even unsanitary conditions. One of the big sources of Mendota pollution is farm runoff—fertilizers and silt from the croplands and barnyards in the upper Yahara watershed. What is the ASCS going to do about it? Thanks to a pilot anti-pollution program, it is going to pay farmers to engage in practices that will cut down on lake contamination. For example, the farmer who doesn't spread manure on frozen fields will be rewarded.

It's a farsighted program, one for which a lot of credit must go to USDA men in Washington and to key local representatives. Now maybe we need a USCS as well—an Urban Stabilization and Conservation Service that will reward city people for not fertilizing their yards or scalping fields to make parking lots.

Congressmen Reuss to the Rescue

Henry Reuss and a guerilla band of fellow Congressmen are taking on the whole U.S. Army Corps of Engineers, and the resulting battle may resemble Bull Run and Buna combined. The Corps has played a noble role in the story of America. Household names like Washington, Lewis, Clark, Fremont, Powell, Warren, Mead, Sherman, and Marshall were Army Engineers. The Corps has built channels, canals, harbors, dams, dikes, locks —literally changing the face of the country in the name of economic development—and pork-barrel politics.

But the Corps has been strangely reluctant to practice on the domestic landscape the same kind of environmental controls it would perform as a matter of course on a military battlefield. For example, since 1899 under the federal Refuse Act the Corps has had the power to ban from all navigable waters and tributaries "any refuse matter of any kind" except liquid sewage. But the Corps has chosen virtually to ignore this mission. Incredibly, the Corps has compelled conservationists to prove their complaint against polluters, but hasn't required the purveyors of sludge and garbage to show their dumping was harmless. All along, the Corps could also have made developers and others give data on the "effluent to be discharged from the proposed sewer outfall," but it didn't bother. It has studiously ignored the pleas of the Fish and Wildlife Service.

Now along comes Congressman Reuss and his Special House Sub-Committee on Conservation. "The Corps of Engineers must make an about-face in handling applications for new landfills, dredging, and other work in navigable water," says the Committee. The Corps should "arrest and take into custody" any offenders. It should request the attorney general to "institute injunction suits." It should eliminate "the substantial gap between promises and performances" on the conservation front.

An Army Engineer lieutenant by the name of Robert E. Lee once came to the Northwest Territory to lay out a military road, and another Engineer lieutenant by the name of Jefferson Davis came to take Chief Blackhawk into custody. They were acting under orders. One thing you can say for an Army Engineer: he should be able to recognize an order when he hears one. The Corps now has a new set of orders—from a Congressman: "Enforce the 1899 statutes." It will be interesting to watch the response.

Conservation Is Luck and Laymen

In a day of "environmental design," "regional planning," and "wildlife management," it may be disconcerting to contemplate how frequently it is that pure chance plays the dominant role in the world of nature and its conservation. But such is indeed the case.

Take, for example, the yews in my front yard. When I bought the place last summer, they were magnificent ten-year old specimens forming a clipped box hedge, sporting a deep green color, without so much as a branch or needle out of place. They must have survived a decade of sun and storm in classic shape. Now they look like something on the landscape of the moon—broken limbs in disarray, browned fronds, sway-backed configuration— which means that something that hadn't happened for ten years has happened this winter. What it was was just the right—or wrong—combination of a night of freezing rain, followed immediately by a day of driving snow, all ganging up to produce a clinging weight on the top of my evergreens that has caused them to bend and snap in obeisance to a natural cataclysm.

Just so in the lifespan of many plant and animal species, it is the fortuitous combination of favorable or unfavorable factors that can make the difference between health and hazard, abundance and collapse. Sometimes chance can work in strange ways. The manager of the federal refuge at Horicon pointed out the other day that the great marsh probably could only have been saved at one particular era in history. It was early in the century, you remember, when Horicon was ditched and drained by hungry agriculturalists. But their techniques were faulty, their equipment crude, their farming practices and produce ill-adapted to muck. So the bubble burst, and Horicon reverted to an expanse of Canada thistles. By the late 1920s and early 1930s, such land was a drug on the market. The conservationists operating at the time had a decent chance to earmark insignificant funds with which to restore Horicon as a duck and goose haven.

But it probably couldn't happen today, biologist Bob Personious points out. Were Horicon not securely in the hands of federal and state wildlife agencies, modern agricultural technology could turn it into productive farmland, and the laws of economics would hence render its acquisition and development for outdoor recreation extremely expensive if not impossible. It may or may not have been chance, however, that there were on the scene at Horicon at the crucial moment men like the Izaak Walton League's "Curly" Radke, who saw an opportunity and seized it. Leadership in conservation may be fortuitous, or it may be the hand of the Great Planner at work.

The Case of the Precarious Popple

The Wisconsin Public Service Commission was in the throes of making a decision that could set a pattern for the wise use of natural resources for decades to come.

The problem before the Commission could be simply stated:

Should we, or should we not, approve the construction of a dam on the Popple River in the northeast corner of the state?

Would that the problem could be as simply resolved as stated. The situation was roughly this:

A group of Milwaukee real estate promoters had purchased or leased extensive land holdings in Florence County and proposed to create a flowage by backing up the Popple. The resulting man-made lake, they said, would offer fine fishing and boating, and the resulting lakeshore cottages would bring $2 million into the county in two years. Arrayed behind the Elco Corporation and Aspen Lake was echelon after echelon of tax-hungry government, from the local town boards to the state capitol, all of them bent on "resource development."

Against this lobby there was only the still, small voice of the
Wisconsin Conservation Department and its coterie of sportsmen-
naturalists, trying to equate the long-term value of wilderness
preserves with the quick returns from commercialization. This
side pointed out that the part of the Popple to be flooded con-
stituted one of the most attractive remaining stretches of wild
river in the state. Arthur Oehmcke of the Natural Resources Department called the Popple "among the top 25 percent of all
trout streams in northeastern Wisconsin." Its white-water rapids
have long been a favorite with canoeists, sightseers, and campers.
What's more, they pointed out, northern Wisconsin needs a new
lake like it needs a hole in the head. There are 172 natural lakes
in Florence County alone. Once inundated, the Popple and its
part of the Nicolet Forest would be gone forever. As they say,
lakes can be made by you and me, but only God can make a
trout stream.

Under Wisconsin Statutes, the PSC cannot permit the construction of a dam that "materially obstructs existing navigation
or violates other public rights." A number of Wisconsin court
decisions have decreed that public rights include "the recreational use of the waters of the state." Were the Popple con-
troversy a traditional case of sportsmen versus power companies,
the PSC would have had abundant precedent for denying the
dam. But here you have a situation where one type of recreation-seeker is opposed by another type of recreation-seeker. Where
do you draw the line between continued exploitation of our
wilderness areas and the preservation of natural domain? How
do you measure the comparative value of ten miles of white
water versus 100 acres of cottages? The PSC voted for white
water.

Were the Popple River dam an isolated case, its significance
could perhaps be discounted. But in a larger sense the problem
of the Popple stands for all the ways in which America's incomparable outdoors is threatened with extinction by the growing
forces of commercial expansion and mushrooming population.
The real solution, of course, ranges far beyond the case of the
precarious Popple. We must move fast and surely toward a series
of land-use formulae for all the country, with certain areas earmarked for more and bigger resorts, and other areas clearly

designated as wilderness preserves. Such a program will inevitably involve the purchase of large tracts of public land. The quicker we act, the cheaper and better the result.

One of the myths of civilization is that every river needs more people, and all people need more facilities, and hence more "developments." The good life has come to depend on the indefinite extension of this chain of logic. That the good life on a river may likewise depend on the perception of its music, and the preservation of some music to perceive, is a form of progress a master plan is in a position to promote.

The Spoilers of the '70s

Look for a legislative proposal to regulate the use of powerboats on lakes and rivers. It will get a lot of support from rowboaters and canoeists.

The situation was brought strikingly home to me, at least, this past summer when I ventured forth to cast for bass as of yore in my boyhood haunts on Rock Lake. The long and the short of it was, it was impossible to fish, or even just to sit and watch the sun set over Korth's barn. The speedboaters kept up such a rock-and-rolling that I retired to Eddie Dettman's spa in fear and disgust. It was not so long ago that the powerboats on Rock Lake were few and far between. Today it seems like everybody north of Rockford maintains a 30-horse power cruiser. There are 10,000 Rock Lakes in Wisconsin, most of them in the same fix.

Lakes, of course, are public property, and everybody is free to use them just about as he sees fit. The trouble is that the power-boaters are ruining a precious outdoor resource for us oarsmen just as surely as if they were siphoning off the water. With raucous shouts and gasoline reeks they deny water recreation to anybody who doesn't measure pleasure in miles-per-hour. A hundred years ago the lumber barons decimated another resource with their "cut and run" tactics, and it has taken us a century to begin to regain lost ground with an expensive reforestation program. Unless a halt is called to the onslaught of a new generation of spoilers, our lakes and rivers will cease to exist as other than liquid truck routes.

One practical solution is water zoning. Land zoning is accepted practice in the cities and the open country alike. Why not zone our lakes and rivers? Under such a plan, Lake A would be reserved for rowboats and canoes, and Lake B would be thrown open to the horsepower addicts. Which lake is which will, of course, call for the judgement of a Solomon, and we are apt to experience a period of stress while the zoning regulations are refined; but the program is essential if our invaluable recreational resources are to be rationed for all types of consumers.

My county board wouldn't have to wait for a state zoning act. It's in a position right now to reserve a bit of water for the rowboaters. All it has to do is continue to refuse to dig the Yahara River between Lakes Waubesa and Kegonsa. Thanks to a thoughtful glacier, this stretch of the Yahara is too shallow for propellers, so it is a natural retreat for muskrats, mallards, and unmotorized man. But the powerboaters are pressing for a massive dredging project. They aren't satisfied with five lakes. They want to turn the whole Yahara into one long speedway. If they succeed, there will be no place left around Madison where you can get away from cruiser jockeys and water-skiers.

As Aldo Leopold once wrote, "twenty centuries of 'progress' have brought the average citizen a vote, a national anthem, a Ford, a bank account, and a high opinion of himself, but not the capacity to live in high density without befouling and denuding his environment, nor a conviction that such capacity, rather than such density, is the true test of whether he is civilized." A practical program of water zoning may be one of the means of developing a culture that will meet this test.

Zeroing In on the Freeloaders

The woods are full of outdoor freeloaders, and the perennial issue before a state legislature is simply this: "Are we, or are we not, going to ask these spongers to start picking up their fair share of the conservation tab?"

It's easy to identify a conservation freeloader, especially when you look him squarely in the eye, like in the bathroom mirror. The list of outdoor spongers is long and varied.

There is the non-hunting bird lover who never buys a duck stamp, although the stamp provides money for marshlands. There is the wildflower fan who never buys a fishing license, although the license helps pay for the acquisition and management of woodlands. There is the outdoor picture bug who spends hundreds of dollars on cameras and not a cent on scenery. There is the camper who gets a cheap ride in state areas supported by sportsmen's licenses. There is the non-fishing water skier who gets by for a $1-a-year boat fee. And there are berry pickers, skin divers, mushroom hunters, nut pickers, hikers, picnickers, sight-seers, all of them coasting along on the conservation program somebody else is paying for.

The historic sources of conservation department revenue are sale of hunting and fishing licenses and federal funds that come from taxes on fishing tackle and guns. Time was when these revenues were sufficient to carry the demands of the freeloaders. But no more. Population pressures are too great.

So legislatures are being asked to spread the costs of conservation more fairly among all consumers of outdoor resources. There is a plan to assess park visitors a higher car-sticker fee. There is another plan to tap automobile license funds, another to dip into gas-tax revenues, another to add an additional cent to the state cigarette tax, another to raise the tax on beer. On the horizon may be a "land use license" for all those who use state land for any purpose other than hunting and fishing. Such a tax has already been proposed in Michigan, where "large numbers of outdoor-loving citizens are not footing their share of the bill," in the words of Michigan Conservation Director Gerald Eddy.

If Eddy is talking about you, you don't have to wait for your legislature to lower the boom. You can voluntarily start paying your keep in the outdoors. Here's how:

1. You can tell your county board supervisor you want some of your tax money earmarked for public-access areas on lakes and rivers in your county.

2. You can help a farmer plant wildlife food and cover patches on portions of his acres held idle under the federal feed grain program.

3. You can make a donation to such private movements as the Prairie Chicken Foundation or the Nature Conservancy.

4. You can buy a federal duck stamp and a state sportsmen's license even though you don't hunt or fish.

5. You can tell your Congressman to help strengthen the pollution-control and wilderness-area bills in Washington.

6. You can tell your state legislator you'll be happy to help finance a state budget big enough to preserve the outdoor resources that will otherwise vanish before our eyes.

New Look for Our Roadsides

As you have driven through the countryside recently, you may have noticed that selected sites on our rural roads are assuming a different look as a new right-of-way management program catches on. This program—called selective brush management—promises to increase the aesthetic quality of country roads, make management of roadside right-of-ways easier, help reduce soil erosion, and provide wildlife food and cover. What's more, these added benefits come at a cost that is less than conventional management.

Yesterday's management practices consisted of stripping all brush and maintaining roadsides in grass cover in the interests of snow removal and visibility. It seemed simple and efficient, but it wasn't. Mowing costs were high. Weeds were a problem. Resprouting trees required periodic treatment. Above all, the loss of a diverse roadside cover for aesthetic, wildlife, and other conservation values along many country roads has proved to be a high price to pay for clean-cutting.

Beginning in 1958, a consortium of state agencies began to experiment with alternate methods of managing our roadsides. What they have developed, in basic terms, consists of removing from right-of-ways tall trees, undesirable woody shrubs, and noxious weeds, while leaving other desirable shrubs and plants like dogwood, hazel, sumac, nannyberry, rose, bittersweet, and grapes. As this "good" vegetation thrives, it eventually limits trees and other "bad" vegetation from establishing itself. The "good" shrubs, vines, and grasses hold the soil, attract wildlife, present a pleasing aspect to the motorist, and provide an effective crash barrier. Gone are the trees that threaten cars and power lines, and weeds like ragweed and thistle. Maintenance costs are reduced.

Selective brush management is not suitable for every stretch of roadside. Its greatest application will be found on county and town roads where travel is light and where all native shrubs have not already been removed. Here, where local units of government have to maintain right-of-ways anyway, often at considerable cost, selective brush management doesn't require any more money and offers the additional values of a diverse landscape plus habitat for wildlife and beneficial insects.

Protest Comes to Conservation

This is the day of the protester, the activist, in a new, volatile, vigorous sense, and it was only a question of time before the tactics entered the world of conservation. The Citizens Natural Resources Association of Wisconsin was up in arms about the continued use of DDT for control of Dutch elm disease in the state, but instead of just holding sob sessions and writing letters, they took somebody into court about it. First, they raised $25,-000 with which to bring the famous new Environmental Defense Fund, Inc. into Wisconsin.

The EDF is a very interesting organization, indeed. It was formed in a New York suburb by a handful of professional people who had not been affiliated with conventional conservation groups but who were darned mad at the continued degradation of soil, water, air, and wildlife. The aim of the non-profit, public-benefit, membership organization is "to prevent serious, permanent, and irreparable damage to the natural resources of the U.S. on behalf of the general public." It doesn't beat around the bush. It goes directly to the courts on serious threats to citizens' rights to a healthy environment.

By way of background to the Wisconsin story, a declaration of mutual intent to prevent the further contamination of the Lake Michigan watershed by DDT and other persistent pesticides had been signed by the heads of the conservation agencies in Michigan, Indiana, Illinois, and Wisconsin. But the Wisconsin Department of Agriculture, which controls DDT use and wasn't

a party to the agreement, had failed to delete from its handbook
the use of DDT for Dutch elm control. This despite the fact
that pesticide contamination has been brought into sharp focus
by the finding that DDT was most probably the cause of death
of nearly one million coho salmon fry hatched from eggs taken
from Lake Michigan brood stock.

Yesterday the tactic of the anti-pesticide people would have
been to write pamphlets, draft petitions, and work through legislative and administrative channels to bring about changes in
spraying codes. Now along comes the EDF with an approach
borrowed right from the yippies. They say there are abundant
constitutional and case law doctrines requiring the protecting
of human life, liberty, and the pursuit of happiness, and they
are taking into court the key public officials involved. It is one
thing for a municipality to debate the use of DDT in a council
meeting. It is quite another thing for the city to face a stiff contest in the courts. If the record of the EDF to date is any indication, active protest is a potent new tool in the hands of the
conservationist.

If you have ever wondered what it was like when David took
on Goliath, you could get a pretty good idea by eavesdropping
on the EDF proceedings.

Goliath in this case was the Chemical Pesticide Establishment,
the whole complicated community of people and programs that
have as a common denominator the mass production and use
of pesticides in general and DDT in particular. As of old, Goliath
came into the fray with a full complement of armor. Unquestionably, pesticides are an integral part of far-reaching modifications leading to major improvements in the efficiency and
effectiveness of American agriculture and forestry, improvements that contribute to the well-being of every citizen.

Arrayed against Goliath in the case was a jerry-built alliance
of fishermen, birdwatchers, and ecologists formed into the new
David on the conservation front—the Environmental Defense
Fund. Compared to the economic armor of Goliath, the EDF
had only an aesthetic slingshot. But the stone aimed straight at
Goliath's forehead was a potent one: "Pesticides are biocides.
They act on fundamental metabolic systems which are common
to many living things; and when they are introduced into an

environment, there is no way of being certain what, in addition to pests, they will poison."

At stake was the question of whether or not the State Natural Resources Board, under various legislative mandates, should declare DDT and DDE to be pollutants of the state's public waters, and ban or restrict their use. Professor Robert Rudd predicted the course of events in his classic book, *Pesticides and the Living Landscape,* a couple of years ago:

> We can expect to see increasing public resistance to programs depending on indiscriminate application of toxic chemicals. Contamination of the environment will continue to mount and will be found to be serious at sites not now suspected. New routes of harm will be discovered, and those now only suspected will be established as serious. Increasing evidence will support a resistant public which, if only intuitively, will demand of scientists and public servants more sophisticated solutions to complex problems and wider responsibility for their actions than now characterize pest control.

Some of the most damaging testimony to the continued use of DDT was given to the hearing on pesticides by Joe Hickey, University of Wisconsin professor of wildlife ecology. "The use of persisting insecticides such as DDT on our Wisconsin landscape is dead wrong," concludes Professor Hickey after ten years of research that reads like a detective story. He says:

1. At least one of the chlorinated hydrocarbon insecticides (DDT) is not breaking down once it is applied. And it or one of its derivatives (DDE) may be transported enormous distances from the point of application.

2. These poisons are concentrating in animal tissues at the tops of food chains in certain interrelated animal systems, wiping out, in conjunction with other environmental forces, entire populations of some bird species in some regions.

In unraveling the evidence that connects DDT to bird mortality, Hickey and others began to worry when predatory birds such as the osprey, peregrine falcon, and bald eagle began having reproductive failures about 20 years ago. It now appears that even small amounts of DDE alter male and female sex hormones. These hormones, along with Vitamin A, are necessary for the addition of calcium to egg shells in a bird's body. Birds containing high levels of DDE are unable to lay eggs with shells of normal thickness. With thin shells, birds accidentally break their eggs and then proceed to eat them. This peculiar egg breakage set in about 1947—a year or two after the widespread use of DDT. For example, the University secured fresh eggs from five herring gull colonies suspected of having different exposures to DDT and DDE. When shell thickness and weight were related to DDE content of the eggs, an extremely close relationship resulted. As DDE increased, shell thickness and weight decreased.

"Statistical tests showed that the relationship we found could have occurred by chance only one time in a thousand," Hickey says. "We regard this field test as a critical verification of the hypothesis that great bird population changes are importantly due to the pollution of our environment with DDE."

The EDF won.

9. The Back-to-the-Land Boom

THERE IS A QUIET REVOLUTION TAKING PLACE ON THE LANDSCAPE, characterized by more and more city folk fleeing to the countryside to stake out little private preserves.

Iowa County, Wisconsin, is typical of what is happening. When I bought my Town of Arena woodlot and put up a cabin only ten years ago, I was one of only a handful of city slickers in the area. Now I am surrounded by expatriots. Further to the west, in the Twin Parks Watershed between Governor Dodge and Tower Hill, the density of urbanites is even heavier. Other clusters of "second homes" are growing near Blue Mounds, Avoca, and Bear Valley. What is going on in Iowa County is duplicated in almost every county within a hundred miles of Madison. The Montello lakes, the Baraboo hills, the McCann valley, the New Glarus glades, Waukesha county, and other areas are feeling the effects of the silent sprawl of the city.

The results are mixed. More and more "No Trespassing" signs are taking lands out of the domain of hunters and fishermen. Land prices have skyrocketed; rural estate salesmen never had it so good. For example, an Iowa County forty that wouldn't have brought $18 an acre a decade ago sold for $180 an acre last week. Carelessly planned developments have turned trout streams into slums. Lonesome haunts of ducks now echo to the roar of motors. Raw, red roads now slash through hillsides once reserved for partridge. Deer try to thread their way through cottage yards where once there were sloughs. Carpets of shooting star and violet are now ordinary lawns.

On the other hand, those adults and youngsters who now have a place where they can "get away from it all" are discovering a new kinship with the out-of-doors, a new sense of stewardship, and a new way to work up a sweat. Eroded croplands are being diverted to grass. Evergreen plantations are restoring a touch of green to sand barrens. Stream banks are being shored up, ponds created, and springs made to bubble again. Disheveled farm buildings sport coats of paint. Birds and sunsets are being watched by people who never saw them before. Pump handles are being worked by business executives, and floors swept with a broom by society matrons.

It is all a part of our new affluence. Where once we dreamed of two cars for every garage, now we must have two garages for every car. It is quite possible that the man we elect as President in 1996 will have been born in a log cabin—a prefabricated cabin, that is. Nobody can predict the net result of the back-to-the-land boom. It may contribute to a dull, gray homogenization of life, with the line between city living and a country retreat so thin as to be meaningless. Hopefully, it can be the makings of a man-land ethic, under which significant numbers of Americans practice a decent respect for our natural heritage.

There's Something About a Cabin in the Country

President Johnson said we should forget about "the good old days" and "count our blessings." While this is two clichés in one sentence, he had a point. One blessing certainly worth counting is a cabin in the country. It means you no longer have to choose between city and country life. You can have both on a modest scale. What was available only to the landed gentry at the turn of the century is now within the grasp of the American middle class. Fifty miles or less from many of our main population centers, much land is empty, beautiful, and for sale. As a matter of fact, the more rural people leave for the cities, the more places they leave behind for city folk who have the money to fix them up.

A cabin in the country is the key to a number of blessings. Men built cities in the first place for safety. They wanted to get away from the danger of remote places. Now the lonely countryside is safer in some ways than the cities. People went to the cities also for conveniences and services. Now, at least in rural Wisconsin, it is far easier to get a man to fix your well pump than it is to get a plumber in an emergency in town. Call him on the party line, tell him your problem, and he comes. He visits and tells you about his own troubles, but he does the job. Above all, a cabin in the country represents privacy and beauty— the two most precious things parents can lend to their children in this distracted age.

"A nation," John Burroughs observed, "always begins to rot first in its great cities, is indeed perhaps always rotting there, and is saved only by the antiseptic virtues of fresh supplies of country blood." America may have indeed reached this point in its story. Philosophically we need the country. Commercially, country land—ten acres or so, a cabin, and a spring that is both bold and true—is the best buy around. Capital, cattle, cheap fertilizer, barbed wire, and flagstone patios have transformed open spaces. But there are still some hillsides, coves, and valleys ready to serve as private parks. They are, as Lyndon Johnson said, one of the blessings worth counting.

In Praise of Country Drama

Some of my outdoor friends think it's sacrilege, but I have to confess I have electricity in my cabin.

A kerosene lamp has a certain charm, and I keep one on the shelf for atmosphere. I occasionally stoke up my wood stove for the same reason. Yet for really efficient lighting, heating, and cooking it's hard to beat Mr. Edison. So my cabin has all the modern conveniences. My theory is, if you want to have the maximum minutes to enjoy nature outside, don't waste hours nursing nineteenth-century appliances inside. As a matter of fact—and I really hesitate to confess this—I keep a TV set in the country. Now I'll have to admit TV may seem an anachronism in a woodlot cabin. But it has a use. For example, suppose you're grouse hunting and it starts to pelt rain. You can retreat to the cabin and watch the Packers. Or suppose you're trout fishing and the sun burns through the clouds. You can take a break and watch Arnold Palmer at Pebble Beach.

The greatest thing about daytime TV, however, is the soap opera. If I hadn't installed a television set in my cabin I would never have discovered *Search for Yesterday, The Noisy Storm, Rage to Die,* and so on. For minds grasping for coherency in a sea of change, a soap opera is a life-preserver. No matter how occasionally you tune in, you can always catch the drift of the plot. In fact, the story will move along at a pace so ponderous as to be positively tranquilizing in a jet age.

This fall, for instance, I got frozen off a deer stand one afternoon when the temperature dropped precipitously. Retreating to the cabin, I tuned in *Edge of Day* by chance. I was able to identify with the main character immediately because he was a newspaper columnist who was about to be arrested for allegedly shooting the editor of the Sunday edition. As the episode ended the police had just entered the city room to handcuff the hero. That was the last I saw of *Edge of Day* until a week ago. In the midst of doing some cabin cleaning I happened to switch on the TV at the right time. Not only was my hero still in view; he was just now reaching the police station! It was as if time had stood still. Where else but following a soap opera can you get such surcease from hectic cadence.

What is more, no matter what your troubles, they fade into insignificance when you become absorbed in a soap opera plot. Our hero, you will recall, is being arrested on a charge of shooting a newspaper editor. On the jailhouse steps who should he meet but his son. Is the son sympathetic? Not on your life. "It serves you right, father," he says. "You are a heartless man. Why else would you have left your six-year-old sons alone in the house the night my twin brother burned to death!" Our hero stands stunned by bitter memories as he faces the jail door. Fade to a commercial for Canary Cold Cream.

I turn from the TV with utter confidence and courage to repair what is now but a minor leak in the cabin roof.

The Death of a Family Farm

One of my farmer neighbors out in Iowa County held an auction the other day. He is going out of business. So another farm dies, a farm that has been in the same family for four generations. In my township there have been a dozen such demises in the past few years, and there will be more. In my state, over 2,100 farmers quit last year. The figure will be higher this year. What we are witnessing is a quiet revolution in land use.

There are a number of reasons why my neighbor decided to quit. In the first place, Arena township is marginal farmland. The soil is second-rate, the hills steep, the bottomlands narrow. You work more than average to produce less than average. In the second place, the costs of farming have risen astronomically in comparison to milk checks. Hired hands are simply not available at all, and the machinery you have to substitute is extremely expensive. More to the point, the style of farming that would work in Arena township is not acceptable to today's generation. You can't practice grandfather's subsistence farming any more. A world given to status symbols and agricultural bulletins will not let you do your own butchering or churn your own butter. You must pose as a big operator or nothing at all. But the straw that is tipping the scale in more and more cases in Iowa County is the phenomenon of the city dweller out roaming the back roads looking for recreational property. The kind of money he is offering for beat up real estate is very hard for a native to resist.

So short a time ago as 1960, when I helped pioneer the trek to the hills of southern Wisconsin, you could pick up a woodlot for less than $20 an acre. Now a hillside with a rock outcrop and picturesque birches will go for more than $200 an acre. As a matter of fact, wooded back forties are bringing more than cropland. The poorer the farm from a strictly agricultural point of view, the higher the price today.

What all this means to the economy is now only dimly seen. The farmer who calls it quits may find an adequate job in town, or he may not. The city dweller who buys the land may husband its resources, or he may not. The farmer who goes to the

city to build a new life may find himself in the company of
fellow ex-farmers drawing unemployment checks. The city man
who goes to the country to get away from it all may find himself suddenly in a variation of suburbia.

One thing is sure. Our instrumentalities of government are
catching up with this new trend only very slowly. Iowa County,
for example, still does not have a zoning ordinance worthy of
the name. But industry is catching on. You can now order a
cabin from a catalogue and buy a wood range at the corner
hardware.

You Can't Go West Any More

My great-grandfather Jones came to Wisconsin in the 1840s
and took out what they called a patent on a farmstead in Wyoming Valley along what is now State Highway 23 between
Spring Green and Dodgeville in Iowa County. It was not on
the main routes of trade and it was marginal farmland, so, with
all of Wisconsin to choose from, you might wonder why Great-
grandfather Jones picked Wyoming Valley. One reason was because he was poor, and Wyoming Valley land was cheap. Another was because he was something of a plunger, and there was
talk about rich veins of lead underlying all of Iowa County.
And a third reason was because he was escaping from the war-
rens of New Jersey, and there weren't many lights in Wyoming
Valley. (In all of these characteristics, Great-grandfather Jones
was probably pretty typical of many of his fellow Wisconsin
pioneers. After all, our ancestors weren't generally the pillars of
eastern society. They were more apt to be the ne'er-do-wells,
the drifters, the dissenters.)

Great-grandfather Jones didn't stay in Wyoming Valley very
long. He planted a few crops and a few sons, and then the
wanderlust caught up with him once more. He headed west in
search of open space and gold, never to be heard from again.
To his descendents, Great-grandfather Jones didn't leave much
of anything in the way of material wealth, but some of us did
inherit his aversion to crowds, which is one reason why, I guess,

I frequently retreat from Madison to my woodlot cabin not far from the original Jones homestead in Iowa County.

My trouble is there are a lot of Jones boys around today, and they're all retreating to Iowa County. The members of the Madison Press Club got a preview the other night of what is in store for Wyoming Valley, thanks to a mammoth "recreation development" funded with Johnson Wax money. Already constructed is a restaurant, the only one ever designed by Frank Lloyd Wright. Now under construction is an 18-hole championship golf course designed by Robert Trent Jones. To be started soon are ski trails, ski lifts, and a ski lodge. Also to be included in the 4,000-acre multi-million-dollar sports, resort, and residential development are a marina, motels, a 300-room convention hotel, riding stables, tennis courts, permanent residences, weekend cabins, a shopping center, and even light industry. In the words of the principal entrepreneur, "Bud" Keland, who also owns the Miami Dolphins, the project "will bring to reality many of Frank Lloyd Wright's concepts of a 'Broadacre City,' a decentralized community that will preserve and enhance the natural beauty of the Taliesin area." To fill up this new city, Keland points out there will be nine million people within 150 miles of Wyoming Valley by 1980.

When things got too crowded in Wyoming Valley for my Great-grandfather Jones a 100 years ago or so, he picked up and went to California. Where do I go today?

Needed: Rural Zoning Laws

In terms of his relationship to his government and his fellow citizens, the city dweller who also owns some rural property lives in two worlds. The city is the world of the big stick; the country is the world of the big carrot.

For instance, let's say I'm tired of shoveling snow from my city driveway, so I propose to move my garage forward flush with the street. In order to proceed, I have to get a city building permit, but I can't, because the law says I must comply with a 30-foot setback for all structures in a residential district. I can

add a side porch, but only if it does not extend too close to my
lot line. I can plant a tree, provided it isn't a cottonwood. I can
put up an eavestrough, but only if the discharge doesn't en-
croach unnaturally on my neighbor. If for sentimental reasons I
wanted to erect an outdoor bathroom, the city health officer
would be on my trail. If I tore down my house and put up an
apartment, I'd be in real trouble with the zoning authorities.
Or let's say I hunger for fresh eggs, and install a couple of am-
bitious hens in a back-yard pen. The police come and confiscate
the chickens, in compliance with a city ordinance. I am allowed
to keep a dog, but only if he lives in a pen and never barks.

I'm not complaining, you understand. I appreciate that all
such laws are designed to protect the integrity of urban areas;
that is, to preserve the equity of the community in what is otherwise thought of as private property. The net effect, of course,
is to make a city man's home something less than his castle.

Now take the situation with my 60 acres in Arena township.
Here I am lord of the manor. I can chop down trees at will,
graze meadows to a nub, plug up springs, contaminate streams,
erect shacks any place, bulldoze ditches, dynamite cliffs, and
otherwise play havoc with the landscape. To try to keep me
reasonably civic-minded, the federal government dangles various
payments in front of my nose in exchange for my engaging in
"conservation" practices; but there are virtually no local laws
to assure that I really maintain a decent respect for the quality
of the environment.

Now I am complaining. It is high time that we rural land-
holders cease to be beyond the reach of zoning ordinances. We
are custodians of irreplaceable natural resources in which the
public at large has a heavy stake. I should have no more right to
wreck my woodlot willy-nilly than I have a right to tear up the
sidewalk in front of my house.

The Return of a Native

Back when I bore the title of city editor of the Lake Mills
(Wisconsin) *Leader*, my mentor was Chester Fuller, superin-
tendent of what they call on country weeklies "the back shop."

Off in a corner where the raw light from an unshaded window glared in on rows of dusty type cases, you could always find Chet hunched over in a sawed-off swivel chair, setting type by hand as he had done since his "devil" days many years before. Those "danged" linotype machines were not for him. In the rack before him, within easiest reach, were his favorite type-faces—Cheltenham, Florentine, and Blanchard, for he was a gentleman of the old, ornate school and would have no truck with stream-lined sans-serifs.

Now glancing at a font of type, now peering out over the rims of his reading glasses at the flow of life on Main Street, Chet composed the *Leader's* display ads. With deft movements his right arm would fly over the case, experienced fingers plucking the right letters. His "stick" rested in his left palm. A calloused thumb constantly jogged the line of characters, keeping them on their feet. A paragraph quadded out, Chester would pause to light the pipe that kept his face in a perpetual smile. Whether its bowl was full or empty, he would strike a match on his lead-bespattered trousers and waft the flame about before his face. After pulling meditatively for a minute he would whiff out the match and with a curious snap of his index finger send it spinning end over end into the "hell-box." At appropriate intervals he would untangle his legs from the type-rack pillars to carry a stickful of type over to the stone, shuffling his big feet along in a comical clog-step, meanwhile brushing flies, real or imaginary, off his bald head.

Mondays through Wednesdays Chet taught me how to set type and lock up a form. After the paper was "put to bed," he taught me how to hunt ducks. Chet was a duck hunter first and a printer second, and Lake Mills was a duck hunter's town. With others of the clan we would go out to his shack on Rock Lake Marsh, in season and out, and study the ways of wind, water, and wildfowl. The boat he gave me has long since been retired but his battered decoys I have kept refurbished for 30 years.

Now I have something else—Chet's old shack, thanks to the thoughtfulness of a number of old Lake Mills friends, who maintain through the years a proprietary interest in the ducks of Rock Lake, and who hunts them. Fortunately I qualify as something of a native. The shack's latest owner, Marvin "Pugger" Maasch, turned over the key to me in a small ceremony. Walter Toepel, the town board chairman, is allowing me and the shack to keep squatting on a point of his farmland. Roger Behling has kindly consented to let me use his road. Phil Engsberg and Donnie Stroede and other old friends helped me fix the place up a little. We didn't make it very fancy, of course. As Chet Fuller always believed, a duck shack ought to be as inconspicuous as a muskrat house and as utilitarian as a can opener.

So, now we're in business on Rock Lake Marsh again. It's good to get back. They say there aren't many ducks any more, but it doesn't take many to make a season. A brace of mallards drifting down over the island, a thin line of geese following the tamaracks along the far shore, a bunch of blackheads buzzing the decoys—that's all it takes; that, and the friendship of people who understand a duck hunter.

Love Affair with Lake Mills

It was 35 years ago that there began for me a rather torrid love affair, which has not dimmed despite the passage of time. It is a somewhat complicated affair in that there are two "girls" involved. One is a lake and the other is the small town on its shores—Rock Lake at Lake Mills in Jefferson County.

From hindsight I can see that I reserved my initial infatuation for the lake itself. To a boy reared in the driftless area of southwestern Wisconsin, moving to a lakeshore was a revelation. A lake is a special treasure—sparkling in the summer sunset, mirroring fall foliage, daring under sodden winter skies, surging to life again in spring thaws. A lake can get under your skin. It can even set the tone for an entire community, almost as if a magic spirit of the waters pervades Main Street. It may be that there is nothing really special about Rock Lake except that she is "mine." For a suitor, that is enough. I pay her court every June with a spinning rod and every November with a shotgun. She never breaks a date. True, I have to share her with more people today than I did in 1933, but I don't care so long as they are Lake Mills natives.

Like their lake, Lake Mills people seem special to me. They

have had the grace to reach out to a prodigal with affection and regard. To the itinerant son of an itinerant preacher, that is everything. Lake Mills is a good town, you know what I mean? Not many very famous people have ever come out of it, that I know. But then there aren't any skeletons in the closet, either. Whisking past Lake Mills on the I-highway between Madison and Milwaukee, it's easy to ignore the town, but if you turn off and drive around the picture-postcard square or stop in at Eddie Detmann's spa, you can get something of the flavor of America's backbone.

The earliest tombstones in the cemetery up on the hill read 1840–1850—Atwoods, and Taylors, and Joeckels, and Fargos— same names as around there today, strongminded people that came a long way to be independent. The Creamery Package manufacturing plant is expanding now and a subdivision is inching its way around the lake; but Lake Mills folk still pay a lot of attention to things like birds, and trees, and plants. Country-style, they watch the change of the seasons with deep interest and awe. I suppose you can find in Lake Mills some characters out of Sinclair Lewis—bumbling George Babbitts and frustrated Carol Kennicotts. If you look hard enough, you probably can even find some Peyton Place types. But by and large Lake Mills people defy the small-town stereotype, mostly I think because they're all in love with Rock Lake.

So if I were ever to write a novel about my hometown, nobody would believe the plot. Nobody, that is, except my pintail and pike neighbors on Rock Lake marsh.

Is Rural Land to Become a Dump?

Back in my hometown we used to have a saying: "The Good Lord takes care of fools, drunks, and Lake Mills." Well, he has done it again. Through an act of Providence the Lake Mills area is not to become Milwaukee's dump, at least not for now.

The story goes like this:

Late last fall local real estate people began taking options on marshlands along the Northwestern Road tracks just to the east of the city of Lake Mills. They were offering prime prices for marginal lands, so the farmers were glad to listen. As the story went at the time, a big manufacturing concern was about to put up a new branch plant. Actually the real estate agents were acting for the Northwestern Road and the National Disposal Company of Barrington, Ill., who were planning to contract with Milwaukee to haul that city's solid wastes and junk to Lake Mills and dump it. By the time Lake Mills folks figured out what was really happening, 400 acres on their doorstep were under option. They protested that the city's image as an outdoor recreation center would suffer, and that air and water pollution problems would develop. But the protests wouldn't have done any good, except that the Milwaukee County Board found a slightly cheaper place to do its dumping—in a swamp in Kenosha County.

This isn't a unique local problem; it is a national problem, and it is two-fold. First, cities are about to choke on their own wastes. Science simply hasn't figured out an efficient, economical way to dispose of or recycle all the things we throw away in our disposable-carton culture. Second, our rural towns and counties persist in rocking along without adequate land-use plans and stiff zoning ordinances that will protect them from wildcat developments.

Into this vacuum have stepped the railroads, hungering for payloads. They are hauling Pittsburgh's garbage 60 miles away and dumping it down old mine shafts, for instance. Toronto's junk is going 100 miles north to be strewn over the moraine. So any small town with a railroad track and no zoning ordinance had better look out. It can have a metropolitan dump on its doorstep before it can say, "Spare that marsh."

The Roar of the Crowd

You can't drive a car within a quarter of a mile of my duck shack at Lake Mills. You have to park at Sandy Beach, cut across lots to the railroad track, and tote your gear down the right-of-way, balancing on 469 ties. You might think this would be a mighty secluded spot, and 20 years ago you would have been right. But not today. For all the privacy it enjoys, my shack might as well be sited on I-94. It is located, you see, on the main channel between Rock Lake marsh and Rock Lake proper, so the traffic past my pier is something to see and hear. The roar of the crowd begins at 3:30 A.M. and continues in increasing decibels until 10:30 P.M. There are rowboats, canoes, dinghies, cruisers, skiffs, houseboats—each with its inboard or outboard, or that newest invention of Satan, the inboard-outboard.

According to the Outboard Boating Club of America, more than 40 million Americans are taking to the water in more than eight million pleasure boats this year, and they will spend nearly $3 billion doing so. You can prove it by me. A big hunk of them are on Rock Lake. Sales of outboard motors will approach 500,000 this year, the OBC says, and they will average 30 horsepower apiece. That's a lot of horses stomping around what used to be a peaceful pond. I have no quarrel with powerboats as such. As a matter of fact, I admit to owning a one-lung outboard myself. And some of my best friends are powerboaters. But I do herewith pick a fight with the aquatic hot-rodders and their atrocious manners: the ski jockey who comes near to swamping my skiff as he swirls around at 60 miles an hour; the ponderous sedan that bears away my line and lure; the banjo strummers who drift by late at night just out of airgun range; the hydroplaners who make a racetrack out of the marsh. My biggest gripe I reserve for the powerboater who insists on cruising the marsh in the middle of the duck season; or worse, who anchors practically in the decoys.

Please, Mr. Town Board Chairman or Mr. Conservation Warden, isn't there some way of rationing the time and space on our jammed inland lakes? Couldn't some waters be reserved for honest fishermen—say from 4 to 6 A.M. on every other Tuesday? And couldn't some waters be barred to high-performance engines between October 5th and November 10th? One ray of hope comes from the OBC. "In another 50 years," promises that organization, "people will be able to skip across oceans safely in high-speed 15-footers." They can't leave too soon for me.

The Lead Gander Goes Down

They buried Ralph Seward the other day, and my hometown may never be the same again.

The paper said he was "a prominent real estate and insurance

agent." He was that, I suppose, but business was really his side-
line. His main occupation was building a community. He was
at the center of just about everything that went on in Lake Mills
for 60 years. Ralph was 90 when he died. The obituary said he
was "the oldest member of the Jefferson County Board." Whoever wrote that didn't know Ralph Seward. Ralph was the
youngest member of anything he ever joined. And he joined a
lot.

Ralph thrived on change. He was an authentic pioneer. From
the day he drove a team of mules to homestead in the Dakotas,
to the day he took a fast jet to visit his daughter in Texas, he
was always eager to see what lay on the other side of the next
ridge. He had his eye on the future. He put his chips on young
people. The amount of change a man as old as Ralph has had to
adjust to staggers the mind. When Ralph was born, Grant was
just barely out of the White House and a satellite was a falling
star. But Ralph adjusted with relish. He was one of the first
people in town to buy a gasoline automobile, and the first to
install an electric stove. If he had lived, he would have been
the first to buy a ticket to the moon.

Ralph Seward not only adjusted to change. He made it happen. He was forever building churches, schools, parks, playgrounds, fish hatcheries, roads, sub-divisions, sewage disposal
plants, factories, wildlife areas. You name it, and Ralph Seward
had a hand in it. He voted for every Republican from William
McKinley to Barry Goldwater; yet when the New Dealers looked
around for a man to get the WPA off the ground in Lake Mills,
they had to pick Ralph Seward—and Ralph accepted.

Ralph was a conservationist of the Gifford Pinchot school. He
believed in wise use of resources for the greatest number for
the longest possible time. Sometimes his wise use got a little
heavy-handed, as when he rammed a highway through the
Schoenfeld arboretum; but more often he was the ever-present
watchdog of the health and long life of Rock Lake.

The preacher who presided at Ralph's funeral quoted from
William Cullen Bryant's "To a Waterfowl." It was fitting. Ralph
was a lead gander, a natural-born general, a community catalyst,
always seen winging toward the horizon, never looking back to
"the good old days." The preacher might also have quoted from

Dag Hammarskjöld: "Only life can satisfy the demands of life.
. . . Don't be afraid of yourself. Live your individuality to the
full—but for the good of others. Don't copy others in order to
buy fellowship, or make convention your law instead of living
the righteousness."

This was Ralph Seward. You can say we may not see his kind
again. Ralph would never believe that. For him the frontier was
never behind, always just ahead.

Part IV

LETTERS FROM SAMMY SQUIRREL, LOBBYIST

ORDINARILY, I SUPPOSE, SAMMY SQUIRREL AND HIS MATE, SYLVIA, would have lived out their lives quietly in their woodlot haven near Mount Horeb, Wisconsin. But Sammy got elected by his fish and wildlife chums as a "legislative representative" for the Southern Wisconsin Alliance of Fur, Fin, and Feathers (SWAFFF). So Sammy went to Madison to become a unique sort of lobbyist. Every now and then he would write to Sylvia back home, recounting his impressions and adventures in the world of *homo sapiens*. With Sammy's special permission, we are reproducing some of those letters in this book.

10. Under the Capitol Dome

Dear Sylvia:

Well, here I am in Madison as the lobbyist for SWAFFF—the Southern Wisconsin Alliance of Fur, Fin, and Feathers. Some of the other lobbyists are having trouble finding a good room, but I'm all set in handsome quarters in an old bur oak right on the Capitol grounds. What's more, I'm finding plenty of nuts in Madison.

How are things back in our Mount Horeb woodlot? I'll keep you posted regularly on things down here during the next couple of years, and I'll appreciate an occasional letter from you and our fellow wildlifers.

The Legislature hasn't convened yet, but I'm going to pay a visit to the attorney general tomorrow to protest some strange goings-on. What has happened already is this:

The Conservation Department proposes to print and sell special stamps that must be affixed to the licenses of trout fishermen and pheasant hunters.

Now this strikes me as very unfair. As a matter of fact, I think this plan is a direct violation of the federal Civil Rights Act of 1964, which specifically prohibits discrimination on the basis of race, creed, or color.

Just because we squirrels do not have green backs like a trout, or just because we don't have feathers like a pheasant, shouldn't mean that we can't have a stamp issued in our honor, do you think?

As far as that goes, I suspect this Conservation Department

proposal is down-right subversive. Otherwise why would they cater to Chinese pheasants and German brown trout while ignoring us 100 percent American squirrels?

I'm going to ask the attorney general to investigate this whole business right away. If the Conservation Department needs money, the least it can do is issue stamps for all of us SWAFFF folks. There could be a grouse stamp, a rabbit stamp, a perch stamp, a pike stamp, and so on.

For that matter, why stop there? I'm going to propose a mosquito stamp for campers, a hidden stump stamp for motorboaters, a poison ivy stamp for hikers, and a rose-breasted grosbeak stamp for birdwatchers.

Think of the revenue my plan would produce! Anybody who stepped outdoors would have to have a book-full of stamps, and stamp collectors who never go outdoors would be after the stamps, too, especially if the printer made a little mistake on a sheet or two.

Maybe I could even work out an arrangement with a supermarket so that a sportsman could exchange his trout stamp for a package of frozen halibut.

If I keep getting ideas like this, Sylvia, there's no telling how high I can go here in Madison.

Yours truly,
Sammy Squirrel

Sammy Gets Tangled in Tape

I got a letter the other day from one of our Fur, Fin, and Feathers members—a mallard duck—asking me if I could help put a stop to his water problems. It seems that between draining and polluting he doesn't have much to swim in any more. So I started making the rounds of the Madison bureaus. Believe me, that's an experience.

Somebody told me the Wisconsin Conservation Department was in charge of ducks, so I went there first. They were very nice but they said ducks, being migratory, are federal birds, and referred me to the U.S. warden. He was very nice, too, but he said the drainage was under the Soil Conservation Service and the pollution was under the Public Health Service.

The SCS boys agreed they give farmers technical advice on how to dig ditches, but that the ditches themselves are under the Agricultural Stabilization and Conservation Program. The ASCS people referred me to a farmer. He wasn't home.

I tried the Public Health Service next. They sent me to the State Water Pollution Committee, which referred me to the State Hygiene Laboratory. They said I would need a water sample to be analyzed.

I went back to the Conservation Department. "Let's talk about water levels," I said. Depends on what water and what levels, they explained. There's the Great Lakes Compact Commission, the Commission on Interstate Cooperation, the Portage Levee Commission, and the Water Regulatory Board, not to mention the federal Corps of Engineers and the state Public Service Commission.

That's if you're talking about surface water. If you're talking about ground water, there's the State Department of Agriculture, the Health Department, and the PSC again.

"Maybe we'd better talk about land management," I said. That depends, they told me, on whether it is county land, state land, federal land, or private land. Unless a highway is involved, and then you talk to the Highway Department.

I wanted to know if anybody is doing any planning and coordinating. Well, it seems some state planning is under the Department of Resource Development, which consults an Advisory Committee on State Resource Planning; and some is under the Conservation Department, which consults the Conservation Commission, which consults a Forestry Advisory Committee, a Program Coordination Committee, and the Conservation Congress.

I also got referred to the University of Wisconsin, which conducts research on water. One of the professors out there said they might be able to set up a migratory waterfowl water committee, provided they could get the College of Agriculture and the College of Letters and Science to agree on a chairman.

In the meantime the professor said I should talk to the Mississippi Flyway Council, the Wildlife Management Institute, the Bureau of Outdoor Recreation, the Fish and Wildlife Service, and the Bureau of Reclamation. It turns out they're in Washington.

I did find one state bureau that seemed interested in our problem—the Board for the Preservation of Scientific Areas and Rare Species. The only trouble is, they don't have any money. So I wound up appealing to the State Commission for the Relief of Innocent Persons. It turns out they handle humans only. But all is not lost. There's a bill before the Legislature that would create a new super-bureau—a State Department of Natural Resources. Whether our mallard friend can hold out until I can talk to this bureau, I don't know.

Sammy Squirrel Gives a Sermon

It's a darned good thing I'm down here. Conservation is going to be one of the hot topics in the Legislature this session, and there doesn't seem to be anybody else around representing the Southern Wisconsin Alliance of Fur, Fin, and Feathers.

One of the big questions shaping up seems to be, "To buy or not to buy." Land, that is. The story goes like this:

Our human friends are running out of outdoor living space. You know how you and I are always just one step ahead of a chain saw or a shotgun. Well, people these days are trying to keep one step ahead of a bulldozer.

A couple of years ago, under the leadership of Gaylord Nelson, the state of Wisconsin passed an Outdoor Resources Act under which money from a special cent-a-pack tax on cigarettes went toward acquisition, conservation, and development of parks and preserves.

The plan has been working very well, with the state beginning to make a dent on the problem of getting recreational lands into public ownership before they are all ditched, drained, leveled, or posted.

Some conservationists want to keep on buying up lands at a fast clip against the day when prices will be even higher and choice sites blighted. Other people, however, want to call a halt to land acquisition and devote more funds to developing right now the lands already purchased. Their main argument seems to be that communities can't afford to see lands taken off the tax rolls without tourist dollars coming back into the economy.

As usual in human arguments of this kind, the real issues are getting pretty blurred by personalities and politics. The Democrats generally are in favor of buying land now; the Republicans want to slow down the acquisition program. Oddly enough, the Conservation Department is in favor of the slow-down; the Department of Resource Development, on the other hand, wants to maintain the land-purchase pace.

At the risk of seeming immodest, Sylvia, I think people could learn something from us squirrels in this case. You know how we work like (pardon the expression) beavers all fall, storing up nuts and acorns. We don't stop every so often to eat what we've buried. We keep salting as much food as possible away against those long, cold winter days when there won't be anything to eat except what we've saved.

If I were a human, I'd use every spare dime I could get my hands on to acquire recreational lands while there still are good sites left to acquire.

Well, that's my sermon for the day, Sylvia. I'll write again soon.

Fish Stocking Facts and Fables

Being a lobbyist in Madison for the Southern Wisconsin Alliance of Fur, Fin, and Feathers is a very educational experience. One thing I'm learning is that we squirrels aren't the only "squirrelly" species. Humans do strange things, too.

Take the current argument in the Capitol about fish stocking:

The Wisconsin Conservation Department has long had a program of raising little game fish and planting them in lakes and streams. There is a cold-water (stream) program involving trout, and a warm-water (lake) program involving muskies, walleyes, and northern pike.

Now there is a good deal of scientific evidence that trout stocking at certain times and places is economically feasible and biologically sound. In fact, there wouldn't be much trout fishing in southern Wisconsin at all were it not for the state's stocking program. Southern Wisconsin streams are in such poor shape that they can't produce enough wild trout to support heavy fishing pressure.

On the other hand, there isn't any consistent evidence that pike and muskie stocking pays off, except in special circumstances. Where there is winterkill, or where a lake is rehabilitated through chemical treatment, warm-water stocking makes sense; but nobody has yet demonstrated that the returns justify the cost of mass planting of pike and muskie fingerlings in assorted lakes.

Given these two sets of facts, you might think that the Conservation Department would be stepping up its trout-stocking program, and cutting down on its pike stocking. Just the opposite is true. In the next two years the Department proposes to spend a quarter of a million dollars more on lake "management."

What is more, the Department proposes to add to its revenues by charging some fishermen more money. But the fishermen who will be assessed are not the pike and muskie fishermen; they are the trout fishermen.

None of this makes sense until you understand something about the politics of conservation. The warm-water fish program is designed largely for northern Wisconsin, where resort operators live off of non-resident fishermen, and from whence come noisy legislators who constantly demand lake stocking programs, whether or not biologists support them.

The trout-stocking program, on the other hand, is concentrated mostly in southern Wisconsin, where resident fishermen and their legislators are a strangely silent lot.

When the Joint Conservation Committee of the Legislature considers these matters, I'm going to testify. But they probably won't pay much attention to me. After all, I'm only a "nutty" squirrel.

Sammy Squirrel Gets Arrested

I have a temporary change of address. Instead of residing in Capitol Park, I'm currently behind bars. It's a long story.

As you know, I've been lobbying in Madison as the representative of the Southern Wisconsin Alliance of Fur, Fin, and Feathers. The other day I attended a hearing on the Conservation Department's pheasant management program.

It seems there's a big debate on whether or not to stock pheasants, where, and when. Before going to the hearing I made the mistake of reading a Conservation Department research report, and I kept quoting from it. I wasn't very popular.

For instance, some legislator testified that pheasant stocking never pays. I said that "with various improvements this program should continue to be an important game management tool under certain conditions."

Some sportsman said the pheasant stocking program should be greatly expanded. I said that "Wisconsin's pheasant stocking program is not the ultimate answer to quality hunting and is strictly a put-and-take program."

Another hunter protested about pheasant shooters paying for a special stamp. I pointed out that "the cost of a bird stocked is at least $1.03."

Somebody said what is needed is more swamps and marshes set aside as preserves. I said that "the best pheasant range is two-thirds cultivated land and one-third cover."

A state senator claimed that what should be stocked are brood hens instead of young cocks. I quoted a statement that "because few hens survive, there is no long-term effect on the pheasant population."

About this time, the committee chairman asked just who I was, anyway, and what my credentials were. I told him. He asked me if I had complied with the lobby law. I said, "What law?"

Then he asked what I was being paid, and I said, "Nuts." That did it. I was rushed out of the room and over to the City-County Building. Some judge there found me guilty of language unbecoming to the Capitol, and sentenced me to ten days in a Vilas Park pen.

I don't understand these humans at all. When some general says "nuts," they erect a statue to him in Bastogne, but when I say it they think it's a federal offense. What I really think upsets them, though, was my quoting all that official literature.

Oh, well, I'll be out of here in a couple more days and back at the Capitol, where I certainly won't quote any more Conservation Department bulletins.

Sammy Gets the Word from Washington

The reason you haven't heard from me recently is because I've been in Washington as a special delegate to the National Wildlife Federation convention.

As chance would have it, the keynote speaker was our own Senator Gaylord Nelson. Boy, did he ever lay it on the line:

"Never before in history have we been so highly organized in the field of conservation—and never before have we been destroying our natural resources at such a rapid rate," Senator Nelson declared.

The trouble is, he said, what we have done in this country is to develop a vast network of special-interest conservation. We have lost our broad vision of the public interest, and we have fallen to quarreling over little pieces of it. We are all passionately in favor of conservation—the particular piece of it that interests us. Our energies are divided among innumerable organizations and causes, each with a special concern with its little piece of nature, and an appalling lack of concern with natural resources as a whole.

That's what Nelson told them.

There were a couple of beautiful examples in Wisconsin in recent weeks of what the Senator was talking about.

For one thing, the Wisconsin Farm Bureau was arguing strenuously that funds should be restored to the federal soil conservation budget so that farmers could continue to receive free advice from Soil Conservation Service technicians, not to mention benefit payments for soil conservation practices.

At the very same time, the Farm Bureau was opposing just as strenuously a state bill that would bring a modicum of control to chemical poisons used as pesticides. The Bureau, in other words, is all for conservation of croplands, but unconcerned about the conservation of song and game birds.

In another instance, people from the Fox Valley area were

plumping for bills that would impose stiffer penalties on paper
mills dumping sulphite wastes into nearby rivers; meanwhile,
the same people were opposing a bill that would require pleasure boats to install better toilets. They're all for fighting pollution, it seems, just so it's the other guy's pollution.

What people have to learn, as Senator Nelson said, is that
insects, birds, fish, animals, water, soil, wilderness, trees, and
plants are all part of the same scheme of nature—a kind of
intricately woven fabric. Snip one thread and the whole thing
begins to unravel.

"True conservation," he told the Wildlife Federation, "is the
protection of the public interest in natural resources."

I think what he meant was that the public includes us fur,
fin, and feathers folks.

Things Are Tough All Over

Before I came down here to Madison I had the idea it was
only we wildlife creatures who are running out of living room.
You know what I mean: chainsaws that level woodlots, bulldozers that fill marshes, silt that muddies trout streams, pollution
that smells up lakes, and so on. Since reading some of the bills
presented to the State Legislature, however, I've decided that
things are tough all over. Humans are running out of living
room, too, and they're trying to ration what's left.

Take Bill 103,A, for example. This bill prohibits anchoring of
boats containing built-in sleeping facilities within 200 feet of
privately owned shoreline without the consent of the owner of
the shoreline property. Next week the Amalgamated Boat Owners will probably introduce a bill prohibiting anyone from
occupying a cottage over-night without the permission of the
guy who has his boat anchored off shore.

Bills 84, A and 16, A try to tackle the problem of how, where,
and when you dispose of toilet wastes while afloat. One proposal
requires fishing licenses now of people angling in what are
called "outlying" waters. Another bill increases registration fees
for motorboats. Still another empowers the Conservation Department to establish uniform "rules of the road" for boaters.

Bill 144,S pits the Conservation Commission against cottagers at Devil's Lake. The Commission wants to terminate the leases so the state land can be used by tenters and picnickers. The cottagers don't want to leave.

Bill 33,A strengthens the hand of the Public Service Commission in policing people who want to place "sand blankets" along shorelines to create beaches at the expense of fish habitat. Bill 261,A prohibits the sale of soft drinks and beer in "throw-away" bottles, so that picnickers won't litter beaches with broken glass that swimmers can step on.

As you can see, all these bills seem to indicate just one thing: Wisconsin people are really feeling the pressure of increasing population and decreasing open space, and they're trying to figure out ways to spread recreational resources around for the greatest benefit of the greatest number.

One bill has me puzzled, however, because it doesn't fit the pattern. Bill 74,A increases the appropriation for recreational advertising to half a million dollars annually.

At least that's one thing we squirrels don't have to contend with yet. Nobody is advertising to lure Illinois squirrels north of the border.

Sammy Gets Set to Lobby

I'm getting set to back three bills now before the Legislature. These bills get at the heart of a problem that is becoming crystal-clear to me. Our present governmental structures are simply inadequate to make real progress on the conservation front.

One of the bills is in draft form already. It is Assembly Bill 753. In a nutshell, this bill provides for the conservancy zoning of shorelands on Wisconsin's navigable waters, with the zoning to be accomplished at the county level, immune from veto by town boards.

At the present time, you see, a town board can block a county zoning plan, no matter how enlightened the plan may be. So in too many places we wind up with shoreland lots that are too small, lack of scenic set-back from the water, destruction of natural cover, septic tanks with leaky drainage beds, unimaginative site planning, over-fertilization on surrounding farmlands, and overt pollution.

In the words of one of the bill's authors, Professor Jake Beuscher:

"Such shoreland mis-uses have a marked impact upon water quality and upon watercourse amenities and aesthetics. If it is true, as the Wisconsin courts have repeatedly held, that the state holds navigable water in trust to protect the public interest, then it is quite appropriate to regulate the shore-owner as may be necessary to assure the fulfillment of public needs."

Another bill I'll be backing hasn't been drafted yet. It was proposed by Al Ehly of the Conservation Department the other day:

"Many counties have separate agricultural committees, soil and water district supervisors, forestry committees, park committees, and other committees concerned with outdoor recreation in some capacity. I would propose legislation creating a new all-inclusive county committee on natural resources.

"Such a committee could activate recreation programs, coordinate conservation programs, effect liaison with the Land and Water Conservation Act, and in general be a real force at the local level. It could have an advisory panel made up of the natural resource technicians stationed in the county. The committee might even come to rival the highway committee in popularity and power."

Al's idea sounds good to me. So does the proposal of Professor Fred Clarenbach. As chairman of a sub-committee studying the problems of Lake Mendota, Professor Clarenbach has come to the conclusion that the task of planning for the management and conservation of the water in the Yahara Basin is almost impossible unless we can devise an instrumentality that can coordinate the controls now scattered among numerous agencies of state and local governments.

"There is no authoritative regional planning agency for the basin, and there is no real 'chairman of the board' who can get each of the many agencies to do its proper part at the right time and in the right way," he says.

A possible solution, according to the Clarenbach committee,

would be to set up a new unit of regional government that might be called the Yahara Region Water Authority. This would require an enabling statute or special act by the Legislature.

I'm all for it.

Billboards and Outboards

If my mail is any measure, two of the most vigorous anti-conservation lobbies operating in Madison are the outdoor-advertising and the powerboat crowds.

The philosophy of these groups is simple: "What's good for us is good for everybody." Under this handy banner they are besieging public officials, asking for special favors, and trying to beat down attempts to protect Wisconsin's outdoors from the sights and sounds of billboards and outboards.

The billboard boys took it on the chin recently when Circuit Judge Edwin M. Wilkie of Madison upheld the constitutionality of the 1959 Wisconsin law limiting roadside advertisements along the new interstate highways.

"The state has a perfect right to legislate roadside advertising in the interests of public welfare," Judge Wilkie ruled.

But the billboard boys aren't licked, by any means. They have come right back with a bill that would exempt long stretches of Wisconsin's I-highways from the act. They are also battling a plan to create a three-man commission that would set standards for signs on state highways. And they are opposing another plan to zone the land around major I-highway intersections.

Front organizations for the billboard boys include the Outdoor Advertising Association, the Wisconsin Motel Association, and the Wisconsin Petroleum Council. Spokesmen say billboards contribute to traffic safety by keeping people awake. They also claim billboard laws destroy freedom of speech.

In the opposite corner are the Wisconsin Roadside Council, the Federation of Garden Clubs, the Wisconsin Automobile Association, and the Friends of Our Native Landscape. Their case for regulating billboards on Wisconsin highways seems irrefutable. Federal aid for highways is partially dependent on conformance with the national rule of limiting road signs along freeways and turnpikes. Billboards at crossroads help cause accidents. Above all, billboards destroy scenic beauty in a state in which scenic beauty is an economic and spiritual asset of major proportions.

Humans have mapped out and begun building what will be part of the largest superhighway system in the world. If they keep it free of billboards it can also be one of the most beautiful. The gliding, effortless grace of these vast stretches of concrete sweeping across the countryside has given a new dimension to travel. If the billboards are limited, the countryside will remain clear to view. In constructing a military transportation system, they shall have built at the same time hundreds of miles of parkways.

The powerboat boys are like the billboard boys. They want to usurp all of Wisconsin for their own purposes. They are concentrating on four campaigns:

One, to permit motor trolling for fish without restriction throughout the state;

Two, to hamstring any plans to zone Wisconsin waters;

Three, to defeat any bills that would regulate water-skiing;

Four, to dredge speedways where motorboats now can't navigate.

Principal spokesman for the powerboat interests is the Outboard Motor Boat Club of America, which is the front organization for the manufacturers of outboards.

Madison folks have a ringside seat at one Motor Boat Club campaign. The powerboaters want to dredge the Yahara River between Lakes Waubesa and Kegonsa. Conservationists want to keep this stretch of the Yahara relatively unmolested. The Dane County Board is on the spot.

The Great Capitol Robbery

If a pair of masked gunmen were to enter the state Capitol stealthily tonight and make off with $600,000 from the Wisconsin Conservation Department safe, tomorrow morning's headlines would be big and black. But because a posse of unarmed politicians is staging the raid in broad daylight to the accompaniment of much oratory, nobody is paying any attention.

There are actually two gangs of safe-crackers at work in the Capitol these days. One crowd has its eyes on a "take" of a third of a million dollars annually. That is the amount of the additional subsidy southern Wisconsin taxpayers would fork over to 27 northern counties each year if the forest crop law gets jockeyed.

The plot goes like this: Under Wisconsin's famous five-year-old forest crop law, the state has been paying 20 cents per acre of county forest to northern units of government as a kind of community-chest donation against the day when these cut-over lands could start producing wood products. To date about $9.5 million has been invested.

In turn, the northern counties pay back to the state a "severance tax" whenever forest products are sold off county forest lands. This tax is pegged by statute at 50 percent of the wood crop value at time of sale. To date the counties have reimbursed the state about $2 million, or $7 million less than they've taken in, not even counting all the free forest management assistance they've received.

But certain northern politicians aren't satisfied with this deal. They want to cut the severance tax from 50 percent to 10 percent. The net effect of such an arrangement would take $3.3 million a year out of the state conservation fund and give it to the 27 northern counties involved in the forest crop law.

The Conservation Department is not concerned alone about any loss of revenue. It is even more fearful of the proposal as an entering wedge that could gradually result in the state forfeiting its long-term investment in the 2,200,000 acres of county forests, all of which are presently open to public hunting and fishing. It will be interesting to see who wins the shoot-out, the general public or the posse of northern politicians.

The second gang of Capitol safe-crackers has its eyes on another $260,000 in Conservation Department funds. That's what it will cost the state to pay bounties on fox, coyote, and wildcat again next year.

Conservation Director Lester Voigt calls the bounty "nothing more than a subsidy" to a handful of bounty hunters. He points

out that the whole history of bounties proves they won't do the job of keeping predators under control.

"The expenditure of a quarter of a million dollars on a useless program at a time when vital conservation needs are so pressing is tragic," says Voigt.

At the current rental rate that sum could pay for the leasing of more than a million acres of hunting land, or foot the bill for almost the entire pheasant rearing program in Wisconsin.

A bill now before the Legislature would eliminate bounty payments, but the gang of northern politicians is gunning for it. They go into the fight wearing bullet-proof vests in the form of an old Wisconsin political myth that whatever Lola wants, Lola must get—Lola being any legislator from north of Highway 10.

Sammy Takes a Look at ORAP

How are you doing out there in our woodlot? If the shot and shell flying around are anything like the flak here in Madison, you'd better keep your head down.

The big argument here is over the future of Wisconsin's Outdoor Resources Act Program (ORAP).

As you may recall, under the leadership of then-Governor Gaylord Nelson the 1961 Legislature passed an act placing a tax of one cent on each pack of cigarettes sold in the state. It earmarked the revenues for a variety of conservation programs, mainly the acquisition and development of public recreational lands.

Called ORAP, the program has a price tag of $50 million and a life of ten years. ORAP has met a real Wisconsin need for public outdoor living space, and has provided a pattern for similar legislation at the federal level and in other states.

Now you might think that being against ORAP would be like being against flag, home, and mother, but the fact is a committee of the Legislature is even now considering a resolution that would place a moratorium on all land purchases under ORAP for at least two years until another Legislative committee can investigate the whole program.

It is the public acquisition of private lands that is the nub of the problem. Farmers and cottage dwellers whose lands are earmarked for takeover can get very unhappy. So can resort owners who see public parks as unfair competition. County board members in the north complain that public land acquitition by the state takes too much property off the local tax base. And then there are legislators who like to make political hay by stirring up any animals that are handy.

While one Legislative junta is bent on scuttling ORAP, another group is bent on extending and expanding the program. This is the special task force appointed by the Governor to recommend a future course of action. The task force has come up with a proposal to approximately double the state's annual investment in land acquisition and development through judicious bonding.

Caught in the crossfire are able public servants in the Wisconsin Conservation Division like Al Ehly of parks. They are castigated by some for buying up too much land for public recreation, and condemned by others for not buying enough while the price is right.

From a squirrel's point of view, Sylvia, it would seem like a silly development in Wisconsin conservation if the State Legislature were to pass a resolution in effect calling a halt to finding and storing up nuts of public land that will be needed in the future to meet the outdoor recreation needs of a mounting population.

The present individual and local problems that stem from ORAP are very real, but they are small indeed, compared to the overall decline in the quality of Wisconsin living that would result if our human friends enter the next decade with open spaces in short supply.

Great Society Swamps Sammy

As a lobbyist for the Southern Wisconsin Alliance of Fur, Fin, and Feathers (SWAFFF), I've got to try to keep on my toes about all sorts of government acts and agencies from town board zoning practices to UNESCO wildlife conservation programs. And believe you me, all this is no small job these days. As a matter of fact, the Great Society just about has me swamped.

Let's take one simple problem as an example. Let's say the county of Dane proposes to develop a new outdoor recreation area, and it wants some state and federal assistance. Is there any? You bet there is—enough to cover the whole park knee-deep in red tape.

Under a couple of different acts, the URA of the HHFA can provide grants for open-space acquisition. The HHFA, by the way, is in HUD.

Under the new LWC act, the BOR also makes money available for open spaces. The BOR is in the DI.

If wayside and scenic overlooks are involved, the new HBA is in the picture. This comes under the BPR of the DC.

If the area in question has fish and wildlife value, two old programs can provide funds—DJ and PR under the BSFW of the DI.

The new LAA helps government bodies acquire land for recreation. The LAA is administered under the ACP by the ASCS in cooperation with the SCS—all in the DA.

I could go on. To do the park development work, you might involve the JC run by the OEO, the NYC run by the DL, the WEP run by the HEW, or the APWP run by the DC.

It's not inconceivable that the ARD would be interested; that's a DC venture separate from the RADP in the DA.

You begin to get the federal picture? I haven't even talked about the potential roles of the BLM, the NPS, the BR, Public Law 566, the ERS, the CES, the FS, the REA, the CE, the FSA, the PHS, and the SBA, not to mention the new maze of laws and bureaus involved in water management and pollution abatement.

In state programs we have a cigarette tax providing for outdoor recreation areas. These funds are allocated by the Legislature according to guidelines propounded by an inter-agency committee representing the WCD, the SSWCC, the DRD, the SDPI, the DPW, the HD, the SBH, the DA, and the governor's office.

At the county level there are committees on just about every-thing except outdoor recreation. Town boards can block any plan, anyway.

Reporter magazine sums up the whole program in these words:

"Probably no series of legislative enactments in U.S. history has created more complex administrative problems than those passed under Lyndon Johnson's leadership. They have three things in common:

"Their implementation cuts across existing departmental and agency lines within the federal government; they demand al-most heroic responses from local governments in order to suc-ceed; they require a combination of technical and administrative skills that are critically scarce.

"The critical shortage (in conservation) today is not money, but people to carry out the programs."

How about coming down here and giving us a hand, Sylvia? I think the League of Women Voters could use you. They're getting in the act, too.

11. Up and Down Main Street

Capitol Park

Dear Sylvia:

I wandered past some show windows here in Madison the other day, and what do you know? Once upon a time they used to name new cars after man; now they name them after wildlife. What Detroit is now selling is not so much a car as a personality. I imagine this presents quite a psychic hurdle for our human friends. The car buyer is forced to ask himself whether his outlook is more aptly expressed by beast or by fish or by bird.

If he decides he is a fish type, he must then wrestle with the question of whether he is most nearly in harmony with the Barracuda, the Sting Ray, or the Marlin. If the Marlin seems to express him best, he has more problems, for he must then decide if he will order it with a "four-on-the-floor super sports transmission for the husband with the adventurous wife."

Is his wife adventurous? This would seem to me like a dangerous question to ask a man. It may start him making discreet inquiries in the neighborhood and lead to nasty family scenes.

But suppose the buyer decides that his personality is bestial rather than marine. He must then undergo elaborate analysis to determine if he is more in tune with the swift Impala, the fierce Wildcat, or the wide-track Tiger.

On the other hand, he may feel more affinity for a Skylark or a Thunderbird. Incidentally, did you ever see a Thunderbird in our woodlot?

229

One species is a sort of hybrid called the Dart, which is neither exactly fish nor fowl but something in which you can "fire up that snarl under the hood" and "head out from the herd," plus "run barefoot through the wall-to-wall carpets."

If the choice were strictly among wildlife species, it might not be too difficult to pick a new car, but Detroit doesn't let humans off that easily. There are also some brand names that describe hunters we have known; for instance, the Rebel, the Marauder, the Rogue, the Fury, and the Caprice.

What a simple old country boy does when confronted with these alternatives, I don't know. It's obviously becoming harder and harder to buy a new car without psychiatric help. By offering a greater variety of personalities each year, Detroit is forcing our human friends to decide who they really are, when most of them probably would prefer to forget.

One thing makes me mad, Sylvia, and that is that nobody has yet named a car after us Squirrels. Why do you suppose that is? I've always thought we had a sort of charm about us, not to mention strength of character. I guess the trouble is that humans use the term "squirrelly" to describe somebody who's not all there. Maybe Detroit is reserving that title for itself.

Yours truly,
Sammy Squirrel

Sammy Squirrel Defends a Criminal

You aren't going to believe this, but a lot of old friends of ours here in Madison are fugitives from the law. The name that shows up on the police blotter is "Populus deltoides," but you know them better as cottonwood trees.

Until recently, the cottonwoods were honored members of the Madison community. Now they are criminals, the city council in its wisdom having declared cottonwoods to be enemies of the people.

It seems that cottonwoods insist on doing what comes naturally to cottonwoods; namely, disseminating their seeds in cottony masses over the landscape. In the view of some persons, this apparently constitutes a serious threat to law, order, and lawns. Hence the cottonwood has been condemned, since he can't be expected to learn birth control.

The city has dispatched its assassination squads into every ward, their orders to saw, chop, or poison to death every cottonwood. In the meantime, other squads of city employees are spraying elm trees to keep them alive. The trees are confused, not to mention us squirrels and some citizens.

A lot of us have a special affinity for the cottonwood. He certainly doesn't look like a crook. His massive branches spread evenly upwards to form a graceful, open top. His leaves flutter and rustle in the faintest air current, like the echoes of never-quite-forgotten shouts.

Now that the cottonwood has been placed on the city attorney's list of subversive agents, it will be only logical if the cottonwood's cousins fall heir to guilt by association. Next spring we can confidently expect that quaking aspen, large-toothed aspen, Lombardy poplar, white poplar, silver poplar, balsam poplar, and Balm of Gilead poplar will join the cottonwood on the list of Madison trees to be exterminated.

For that matter, the city council may not stop with the "Populus" family. There's a lot of nuisance vegetation around, if you want to look at it that way.

In the spring, spent honeysuckle, spirea, and forsythia blossoms litter lawns. In the summer, grass insists on encroaching over the edges of sidewalks. In the fall, maple leaves create traffic hazards on city streets. In the winter, we squirrels dig holes searching for acorns that wouldn't be there if they outlawed oak trees. All year 'round, evergreen disturb the peace with their perpetual sighing in the wind.

According to my tree guide, the name "Populus" was given to the cottonwood family because "the music of their fluttering leaves resembles the murmurings of an assemblage of people." Between the murmurings of the city council and the murmuring of the cottonwood trees, some of us will take the cottonwoods.

Sammy Sees the Strange World of TV

In between attending legislative sessions and hearings, Sylvia, I pass a good deal of time watching television. Television is a box with a glass front through which you can watch portrayed all sorts of human doings. By watching TV I have learned a lot of interesting things about people.

For example, I used to think there were many different kinds or types of people, but that isn't so; as least, not on TV.

There seem to be only two kinds of American women. One spends all her waking hours at the sink, enamored of cleansers and making hysterical exclamations about very small matters. The other is gloriously young, beautiful, and wildly flirtatious, and continually swings her long, shining hair back and forth.

Both kinds of women are constantly baking—or heating up— the richest possible cakes and cookies, yet they have figures like those of young girls. The pastries, it seems, are all done for the children. Heaped plates are put before them, and they gobble them up without so much as a thank you.

There are only two kinds of American men. One is a young, handsome bachelor who, unfortunately, is afflicted with dandruff. But he pours a secret elixir on his scalp and goes on to countless conquests.

The other is a husband who comes home very tired from a hard day at the office, and his wife (in the kitchen, naturally) greets him and says not to forget the PTA (whatever that is) meeting that night. He is annoyed and snaps at her, but then he takes a pill, and everything is all right.

I feel right at home watching television, because a lot of our wildlife friends and neighbors are TV stars:

A dove who flies into a kitchen window, excites all the neighbors, and then turns into a detergent.

A bear who turns a beer can into a swimming hole.

A duck who moves a whole houseful of furniture to Arizona.

A collie dog who helps put out forest fires every week.

A porpoise who cavorts in and out of all sorts of situations.

A kitten whose paws are as soft as a baby's diapers, provided the diapers are washed with a certain soap powder.

A lion who announces movies.

An elephant and a donkey who preside at political rallies.

A badger who is a football team mascot.

All of these bird and animal performers are the protégés, they tell me, of men who occupy cubicles in a huge glass slab in New York, where they crouch over typewriters trying to formulate new ways to make 50 million people buy new products.

Some day, maybe, one of those men will get the idea that a squirrel is a natural-born salesman, and then you and I will be on television, Sylvia. I can hardly wait.

Sammy Fills Out a Form

I may be in trouble, Sylvia. I got one of those forms from the census bureau the other day, and had to supply what they call "demographic data." The way I filled the thing out may subject me to a full field investigation by the CIC.

For example, right off the bat they asked me what color I was. I hedged by saying "western fox," which is technically correct, but now I'm wondering if they'll realize that means I sport a red flag for a tail. If they do, I may be pilloried by the Internal Securities Committee. Of course they wanted to know if I was married. What can I say? According to Trippensee's standard text on wildlife, "squirrels are promiscuous in their breeding habits." So that's what I wrote. That will shake them up. Next they wanted to know how big a house I live in. That stumped me. As you know, our den is about two feet by two feet, but during the course of a day we range over two to three acres, and during a season we cruise up to 40 acres. I put down "30 rooms." That ought to stump them.

"Where do you live?" they asked. I said, "In a rural agricultural district on a farm woodlot bordered by crop land." I probably should have added that we need cover lanes connecting with neighboring woodlots and fields. After all, if people need highways, we ought to be able to say we need fencerow travel routes. Then they wanted to know my principal diet. I told them, "Mast." I bet they have to look that one up. You can't find anything called mast in the A&P. "What do you think of the international situation?" they wanted to know. I told them I was much more worried about hawks than about doves, and about domestic shooters with 12-gauge shotguns than about overseas insurgents with Russians rockets. I'm not sure they'll be able to feed that answer into the computer.

"Do you have any thoughts about law and order?" the form asked. Here I quoted Trippensee again: "If a squirrel must be handled manually for longer than a brief moment, it is advisable to anesthetize it first; squirrels are vicious, and their bite is painful." They'll probably think that means I'm voting for George Wallace. The census boys wound up by pumping me about Great Society programs. Here I referred them to our old friend Durward Allen, who once wrote about squirrel management:

"Food and belly fat, shooting and living space, tree dens and protection, healthy squirrels and many litters—that, it seems, is the formula."

I can't find anything about such things in the platform of either party.

Sammy Looks at Violence

It used to be that when you called somebody an animal, you were somehow separating him from the human race. Writers who talked about "nature red in tooth and claw" were not including people. Now, after observing things here in Madison and on TV, I'm beginning to wonder.

This *Homo sapiens* is a pretty violent fellow. Racial warfare is a constant threat, crime is increasing, and the city is at least as dangerous a place as our woodlot. I've watched Chicago police clubbing hippies and yippies, and I've read about other riots, murders, and a random assortment of beatings in the streets. It's all not much different than what takes place in the so-called wild.

Maybe this isn't so strange after all. You can make a case for the view that man is the only carnivorous ape, and that he wouldn't be here as man today if he weren't violent. It's a pretty good bet that at some fateful moment in history a long-extinct ape uncle turned back to the bush to starve on a dwindling supply of berries, while an ape grandfather took to the savanna with murderous intent and a taste for meat.

People are still the only species that casually kill their own kind, and lately they have become fascinated enough by the habit to keep IBM records of the incidences. Americans don't have a corner on violence, although you might think so by listening to some politicians. German massacres of Jews, the strife that accompanied Indian partition, the destructiveness of the last days of French Algeria, mass executions in Russia and Indonesia, civil wars in Africa, fighting in Jordan—it seems pretty clear that the knack of violence is widespread.

From all of this, Sylvia, you might get the idea that human beings are not different than our fellow fur, fin, and feather friends. But I have to admit there is an opposite side to the coin. Humans have a concern for brotherhood that goes beyond mere words. It is expressed today, for instance, in a massive spontaneous effort called "the ecumenical movement" among the churches. There is a growing concern in business, in government, and in the private conscience for the unemployed, for the uneducated, for the ethnic minorities, for the old, the deprived, the brutalized, the homeless, and the disenchanted.

We have to remember these traits when we assess *Homo sapiens*. That species has always been inclined to think the worst of himself—immature, uncivilized, materialistic, violent. I think he would do well to put aside his obsession with his faults, look at himself realistically, count his blessings, and get on with solving his problems.

After all, that's what we animals do.

Sammy's No Leaf-Raker

All over town on spring days human beings are engaged in one of the most curious ceremonies I have ever seen. They are raking leaves off their lawns and consigning them to municipal dumptrucks.

This is a strange ceremony for several reasons, it seems to me. First, the people who so fastidiously manicure their own lawns apparently never give a thought to the fact that their leaves become the trash that scars somebody else's environment somewhere else. Second, the people who remove natural fertilizer in the form of leaves in the fall are the same people who dose their lawns with artificial fertilizer in the spring.

Years ago, they tell me, people either let their leaves accumulate, or they burned them in the backyard garden—the bonfire making a splendid autumn pyre, the pungent smoke penetrating youngsters' nostrils so as to last for years, the ashes restoring nutrients to the soil. But not today. The suburbanite who lets his leaves go untended is socially ostracized. It's even against the law to burn them. To assist him in raking, hauling, and maintaining status, a whole industry has grown up—mulcher attachments for power mowers, wide bamboo rakes from Japan, big plastic bags from Monsanto, and of course the fancy dump trucks and street sweepers, some with vacuum-cleaner attachments.

Thus in another minor yet meaningful way does man interrupt the ecological cycle. Instead of husbanding his tree humus he has it carried off to pollute somebody's watercourse. We can only be happy, Sylvia, that nobody rakes our Mount Horeb woodlot. It may look a bit untidy, but it's healthier than any Madison yard.

Sammy Squirrel Discovers Easter

Today is something called Easter Sunday here in Madison. Easter is a rather curious festival. At least it is difficult for a squirrel to understand.

At first glance, you might think Easter was some sort of a farm produce fair. For weeks, store counters have been full of eggs—eggs of all sizes, colors, and materials. Store windows have been full of little chicks and ducklings, and baby domestic rabbits.

A Chief Rabbit, otherwise known as the Easter Bunny, is supposed to have visited every house in town last night and hidden assorted eggs in assorted places for youngsters to search for this morning.

I thought maybe this was another example of the way the government disposes of surplus agricultural commodities, but the custom turns out to be much older than the farm problem. In fact, it can be traced back to pagan rites at the dawn of time.

From another view, you might think Easter was a fashion parade. This morning all sorts of humans could be seen on the streets, many of them wearing weird hats of many hues.

Interestingly enough, unlike the case with our bird friends, it is not the male of the human species who takes on fancy plumage in the spring; it is the female. She displays herself much in the manner of a cock pheasant, and the drab males cluster around her. I suppose some biological function is served by this performance.

As I have discovered upon further research, however, the true Easter has nothing to do with rabbits or regalia. It is a religious holiday. In fact, it is at the very center of the Christian faith, commemorating as it does the resurrection from the tomb of one called Jesus, and the promise of immortality to those who believe in His name.

This faith is not held by all humans, and even some Christians can be classed as doubters, because it seems to be popular in some circles these days to speculate that "God is dead."

I can understand how such an idea might find currency among city folk. After all, there is nothing about the urban landscape to suggest that there is an order to the universe, and that spring's rebirth follows immutably the hibernation of winter. City life suggests just the opposite—that the disorder of traffic jams is a fundamental law, and that when something dies it is dead as a doornail and nobody cares.

Somebody familiar with our woodlot, on the other hand, would have a hard time *not* believing in a natural order and an eternal life. Our oak tree loses its leaves each fall, but it has never failed to revive each April. It will certainly decay and topple over some day, but only to shelter and fertilize a crop of seedlings.

I don't know whether such an observation makes me a cultist or a saint, and fortunately we squirrels don't have to worry about such labels.

Sammy Looks at Fishing

Spring is the time of the year when a strange malady affects many of our human friends. The symptoms are varied, but they usually consist of things like glassy eyes, sweaty palms, and a twitchy forearm. Victims putter around in basements or sit staring out of office windows. Doctors call the disease "fishing fever." There is no known shot that you can take to ward off this fever, but fortunately there is a sure cure. You get a pole, a hand of line, and a hook, which you dangle in a stream or lake. Within moments the fever calms down. The only problem is, it comes back again, and you have to repeat the cure. You can see hundreds of humans taking the cure along or on any body of water.

Fishing fever is catching. It is spread by word of mouth over backyard fences or around poker tables. Some carriers of the fever are particularly virulent. They are the writers who spin what are called fish stories in magazines and newspapers. Some nurses do not believe in allowing the victim to take the cure. They are the wives who draw up lists of yard work and house-cleaning chores in an attempt to bleed the fever away. There is no evidence that this approach will abate fishing fever, but it will break up families.

America being the kind of society it is, manufacturers have learned how to capitalize on fishing fever. They turn out all manner of anti-fever aids, which are displayed flamboyantly in sporting goods store windows, and on the sprawling counters of discount houses. These aids are known as fishing tackle. They consist of a bewildering array of rods, reels, monofilament lines, lures, tackle boxes, vests, boots, and the boats to haul them in.

No doctor has ever proved that you need all this special dope to fight fishing fever, but that doesn't seem to make any difference. Victims fill their closets and garages with patent medicines nonetheless. On all this tackle they pay a special tax that is used to buy and develop public fishing waters. It's a very strange case of humans financing the fever that attacks them. To cater to people with fishing fever there has grown up an industry called the resort business. They tell me it is one of the biggest businesses in the state.

Maybe we should open up a resort along the creek that runs through our woodlot, Sylvia.

Sammy Takes a Vacation

The reason you haven't heard from me for some time is that I've been on a vacation. A vacation, Sylvia, is a peculiar institution that humans engage in, most frequently during the summer months. According to the dictionary, a vacation is a "surcease from labor." The truth is though, Sylvia, that nobody works so hard as the average American on his vacation.

Take, for example, the typical family tour to northern Wisconsin. There is, first of all, the "prior preparation" stage, in which you collect by dint of laborious correspondence the data on where to go and what to see, culminating in a moment of decision that is probably responsible for more divorces than any other single domestic confrontation. Once the objective is distilled over the fires of family debate, you collect by dint of excruciating commandeering a stockpile of exotic gear that always adds up to one more boxful than you can cram comfortably into your stationwagon. So at the last minute you leave behind a dufflebag that turns out to have contained a crucial camp item like toilet paper. At your appointed "rendezvous with relaxation," as it says in the tourist literature, you devote so many hours to the sheer logistics of maintaining minimal creature comforts, that you have less time to lie in a hammock than if you had stayed in your own back yard. And it will take at least a whole day out of your schedule just to pack up for the return trip, at which time you will again wind up with one carton too many. So you leave behind a priceless shank of driftwood or a genuine Chippewa war bonnet. You do not, however, fail to bring back 27 rolls of film, from which you make slides to edify neighbors and friends. The only trouble is, they will have vacation slides of their own. Thus a winter evening in suburbia is spent in the sheer agony of watching an out-of-focus sun set over Lake Minnewhozite.

Why all the grim determination to make a major expedition

out of even the most minor vacation? It's what they call the
Puritan ethic, Sylvia, under which hard work is Godlike and
idleness is of the devil. Hence you turn your time off into toil
in order to get to heaven dead tired. All in all, Sylvia, I doubt
if I will ever participate in another human-type summer vacation. It's much more satisfying just to sit curled up in the crotch
of our woodlot oak, getting to heaven at last by going all along.

Sammy Reviews Recreation

To us members of the Southern Wisconsin Association of Fur,
Fin, and Feathers, Sylvia, it may be a little hard to realize that
outdoor living has become sort of a Holy Grail for our human
friends. We take pretty much for granted the privilege of sitting
on a woodlot log, or playing tag in a treetop. Not so *Homo sapiens*. After eons devoted to escaping from the veldt, he is now
bent on a headlong return to nature, or what is left of nature.
I joined this twentieth-centry crusade this past summer, and
discovered some interesting things about Outdoor Recreation,
AD 1970s.

First, there is developing a peculiar Outdoor Recreation philosophy—a strange combination of Yankee grit and oriental mys-
ticism. The muscular strain traces back through the fadism of
physical education cultists and the Puritanism of the pioneers to
the Spartanism of the Greeks. The mental strain stems from
eastern and occidental Indian postures as interpreted by the
Thoreaus and Muirs who have seen in nature a vision of man
in tune with the cosmos.

Second, to implement this philosophy, Outdoor Recreation
now represents a growing matrix of government bureaucracy
and private enterprise that may well come to rival the military-
industrial complex. The federal, state, and local agencies devoted
to husbanding outdoor resources are exceeded only by various
forms of commerce devoted to exploiting those resources—all in
the name of Outdoor Recreation.

Third, in every Outdoor Recreation area today we see the
cataclysmic confrontation of a rampant population and a ram-

pant technology versus a fragile, finite resource base. Man has spent enough money to get to the stars, but not enough to handle his sewage. He has learned how to destroy a Vietnamese jungle but not how to protect a Vermont mountain. He treats every stretch of countryside as if he had a spare in his hip pocket. So human Outdoor Recreation today is sort of a dream castle anchored on quicksand. Will our human friends get their priorities straight in time? It will take an earthbound Apollo program of the first magnitude.

Science Comes to Shooting

I toured the sporting goods stores here the other day, and I tell you, when hunting season is open, you're going to have some very fancy lead thrown your way. Science, you see, has come to shooting.

For example, for years gun designers have been trying their best to take the "kick" out of shotgun shooting. It now appears they have succeeded. A new recoil reduction system—utilizing hydraulics and powerful springs—measurably reduces shoulder impact from firing to a gentle, almost sociable, nudge.

By effecting a 78 percent reduction in recoil—or about half the kick of a 20-gauge shotgun—this new device will be a boon particularly to oldsters suffering from ailing joints and to young hunters. According to the manufacturer, "the veteran shooter might also find the new system to be the difference between consistent hits and misses."

In other words, Sylvia, you'll have to dodge a little faster this fall. You'll get some help from another new development—an illuminated tracer load that visually pinpoints the center of a shot charge in flight.

The heart of the new tracer load is a specially designed capsule containing a pyrotechnic element that ignites when the shotshell is discharged. The tracer element is visible to the shooter all the way to the target. According to the label on the box, the shooter is thus provided with a true indication of the trajectory of the shot charge. Instantly he can see why he hit or missed and adjust his next shot accordingly. What it doesn't say on the box, Sylvia, is that *you'll* be able to see that shot charge, too, and duck all the quicker.

Humans have been making firearms of distinction and aiming them at us squirrels for a hundred years now, it says in the papers. It was in 1866 that Oliver Winchester made the first firearm bearing that famous name—a lever action piece. The Indians called it "Yellow Boy" because of its brass receiver, and it was held in great respect due to the efficiency of its 17 shots.

Winchester commemorated its 100th anniversary by issuing a modern centennial descendant of that first lever action. Having the appearance of its famous history-making ancestors, the new lever action will have an octagon barrel, a shiny gold-plated receiver, and a crescent-shaped solid brass butt plate. The walnut stock and forearm have the straight classic lines of the old Model 1866. The sights are the same buckhorn style that frontiersmen used to draw down on buffalo.

Fortunately, Sylvia, you won't have to worry about this fancy new weapon, because it is being made only in 30-30 caliber. At least I haven't heard yet of anybody using a 30-30 on squirrels. You never can tell, though. This is a day when people are trying to get a bigger bang for a buck.

Sammy's Glad He's Free

I know you must think us squirrels live a tough life, trying to raise families in holes in trees, and dodging owls and shot. But, really, things aren't so bad for us. At least we don't have to get a permit to harvest nuts, nobody tells us how many we can take, and we don't need a passbook in order to store them. In contrast, listen, if you will, to the sad fate of Wisconsin duck hunters this year. Their daily bag is reduced to three birds a day. This bag cannot include more than one canvasback or redhead, one mallard, two wood ducks, two black ducks, or one hooded merganser. To go hunting they of course need a state license and a federal stamp. On top of the state license and federal stamp, Canada goose hunters need a special permit. Around Horicon

Marsh this will entitle them to one bird a year. In the rest of the state they can take four if they can find them.

It's all because duck populations are way down, and goose populations are too concentrated at Horicon. The regulations are very necessary, in other words, but they make a pretty regimented business out of what was once the epitome of free America. And it's not only hunters who are regimented these days. They held a "democratic" convention in Chicago the other day, and the policemen outnumbered the delegates.

To understand all this, Sylvia, you have to realize that human beings are really just numbers, or rather lots of numbers. There are social security numbers, draft card numbers, credit card numbers, bank account numbers, and so on. Even students registering at The University of Wisconsin last week were assigned numbers. You can forget your name and carry on, but if you misplace a number, you're sunk. Of course humans have one privilege we squirrels don't have. They can shoot us, but we can't shoot back. To be sure, they need a permit to do it. Oddly enough, they don't need a permit to shoot other people. All told, I'm not sure I'd trade places.

Sammy Sympathizes with Hunters

This is just a friendly warning, Sylvia, that the Wisconsin squirrel season opens Saturday morning, September 30th. You'll be under the gun for four months until the last day of January.

I know it's tough, Sylvia, dodging shot and shell every fall, but it may be some comfort for you to know that the life of the modern hunter isn't any bed of roses.

Take the case of the Wisconsin goose shooter. First, he must buy a state hunting license, to which he must affix a federal migratory bird stamps. Then he must get a special permit to hunt geese. Finally, he must apply for a goose tag to affix to his goose.

The application for permit and tags had to be in the mail to the Wisconsin Conservation Department by August 30th, no later.

If you want to hunt in an 11-county area surrounding Horicon Marsh, you will be allowed to take just one Canada goose, and you will be assigned by computer to hunt in one of seven periods during a 37-day season beginning October 14th. If you are willing to hunt in the balance of the state, you will be allowed to take two geese, and you can shoot any time over an extended season.

To hunt in the best spots in the Horicon zone, you will have to rent a blind on a daily-fee basis from a farmer. There will be no public blinds available adjoining the federal refuge this fall.

If all this sounds like an insidious scheme devised merely to harass hunters, it's not. While the tag system is new to Wisconsin, it has been used with success in other states to limit overkill, distribute hunting opportunities, and improve hunting quality.

Goose hunters aren't the only sportsmen caught in the jaws of button-down-collar conservation. There is no open season at all on bobwhite quail, prairie chicken, or sharptailed grouse. Woodcock shooters are presented with an unprecedented early opening date, but they won't be able to see the birds for the foliage. Duck hunters must thread their way through the usual complicated restrictions on hours, species bag limits, and hunting techniques.

For everybody, there's a new prohibition against loaded or uncased guns on state-owned wildlife areas in six southeastern counties except during the hunting seasons, while training dogs, or at a designated target range.

All told, Sylvia, it's a lot simpler if not safer to be a squirrel these days.

Duck Hunting Roulette

Although you may think differently, it is not true that all the hunters in Wisconsin are tramping your woodlot this weekend. A sizeable number of them are hunting ducks. And if you think duck hunting is all pleasure, guess again. Unless you call playing Russian roulette fun. You see, Sylvia, in order to stay on the right side of the law, a duck hunter has to be able to distinguish mallards, canvasback, redheads, wood ducks, and scaup from all other species. Why? Because there are special limits on these birds. What's more, this identification has to be performed while the bird is on the wing, and often in the uncertain light of early morning.

How well does the typical duck hunter do? Not very well, according to a recent study by biologist Jim Evrard. Jim tested a group of Madison "experts" on Lake Mendota, and they scored no better than 75 percent. A group of novices were only 50–50. A group of average waterfowlers at Horicon scored 70 percent. Yet the regulations assume that everybody in a duck blind can be 100 percent accurate consistently.

There is one bright spot in the Evrard data, however. He was able measurably to improve the ability of duck hunters to identify birds on the wing—through a series of training sessions. And the hunters he talked to said they'd be glad to go to "duck hunter schools." So it may be, Sylvia, that in another year or so our duck hunter friends won't be able to get a license until they can pass an identification test. The only trouble is, those who flunk will probably fall back on squirrel shooting.

Sammy Looks At Deer Hunting

Your woodlot is undoubtedly a crowded place on a November weekend. From dawn to dusk you are watching a parade of red-clad human figures stomp through the underbrush or sit warily on stumps. It must be a great deal of consolation to you that these hunters aren't after you. They're after deer. Oh, one or two of them may take a poke now and then at a squirrel with a sidearm, but the whitetail is their main quarry between now and the 29th. They are part of the 500,000-plus people who take to the woods each November.

These deer hunters may never have thought of it that way, Sylvia, but they are actually performing a profound biological function. They are helping to cut the deer herd down to the carrying capacity of the range. If it weren't for the annual shotgun and rifle harvest, the deer population in Wisconsin would get out of hand. In other words, Sylvia, deer hunters are simply the modern predators who take the place of yesterday's wolves, cougars, and screw worms.

An interesting point, though, is that human beings themselves don't have any natural predators outside of themselves, and homicide is illegal except on the battlefield. Besides that, their medicine is so effective that lots more people are living longer. And they don't practice birth control to speak of. So the human population, they tell me, may be getting out of hand. Depending on whose figures you use, the world may be so crowded in "x" number of years that physical and psychological starvation will begin to operate.

It's simply a case of a basic ecological law working itself out— the law of carrying capacity. So much real estate and resources will support only so many animals. And man is an animal in this sense, just like a deer or a squirrel. Deer hunters clustered around camp stoves might just think about that for a minute. Unless people start limiting human population growth, all other conservation measures may not be worth the candle.

Sammy Observes Thanksgiving

I trust you had as peaceful a Thanksgiving as possible, what with holiday hunters tramping through our woodlot. I had a fine dinner of hickory nuts here. I had hoped to get out to Mount Horeb, but there's just too much doing in Madison on the conservation front. More people are getting concerned about environmental degradation than ever before, and they're trying to do something about it, on scales large and small. I took the occasion of my Thanksgiving dinner to say a prayer for some of the ordinary creatures like us who are beginning to get their backs up in small but significant ways.

For example, there's this voluntary group on Madison's west side that is trying its darnedest to preserve a pothole. They call it Kettle Moraine Park. It's the last bit of undisturbed wetland in a subdivision, and some developers want to use it as the site for a new apartment complex. But the neighbors are saying, "No." Oddly enough they aren't getting any help from city park officials, who apparently can't see the purpose of an odd parcel of nature that isn't supplied with a ball diamond and picnic tables. But the pothole people still are poking away at the problem. They've even raised $5,000.

Then there's another group of people who want to preserve a vintage stone house out on University Avenue. It's the only touch of charm in a stretch of neon jungle, and it's destined to become the site of a hamburger joint. This will be an uphill fight, because zoning ordinances and tax structures are set up in such a way that it's extremely difficult and expensive to conserve graceful pulchritude in the face of "progress." But you have to start somewhere.

I said a prayer, too, for those of the "silent majority" who are writing letters of protest to local TV stations about dishonest snowmobile advertising. The airwaves are full these days of commercials showing different breeds of snow vehicles gliding over drifts and flying through the air with the greatest of ease. The trouble is, these scenes are accompanied by the strains of symphony music instead of by the actual earsplitting noise of the snowmobiles themselves. If the snowmobile manufacturers had to use synchronized sound instead of substituting artificial music, TV viewers would have to leave their livingrooms. Come to think of it, I'm going to write to Vice President Agnew about this. He wants the TV networks to "tell it like it is" these days, so he might as well "T" off on phony commercials instead of just castigating David, Chet, and Walter.

Sammy Looks for Christmas Spirit

This is supposedly the season of good will to men, Sylvia, but from some of the things I've been reading it's hard to work up enthusiasm for the human animal. Take the recent survey on who litters and why, run by Keep America Beautiful, Inc. It shatters a lot of illusions. For example, you might think that the younger generation, with all its new-found concern for the environment, would do a better job of outdoor housekeeping than their elders. Not so. People between the ages of 21 and 35 litter twice as much as those between 35 and 49, and three times as much as people over 50. Or take the idea that litter is a city problem. That isn't so either. Farmers and residents of small communities litter more than people from large urban areas.

The angling fraternity doesn't come off very well in what I've been reading either. During a period from May 23rd to July 7th, Canadian conservation officers checked 8,951 fishermen in the Port Arthur area. They found that a staggering percentage are meat hunters first and true sportsmen second. Over 500 fishermen were warned for various infractions of the fishing laws, and 262 were charged with violations. The officers were obliged to confiscate—get this—2¼ TONS of illegal walleyes. This isn't angling, it's plundering.

But all isn't dark, Sylvia. I've also been reading about the walk staged on a rainy day one October by devoted conservationists through the Sylvania Tract of the Ottawa National Forest near Watersmeet, Michigan, protesting the way in which the U.S. Forest Service seems to be turning this vest-pocket wilderness into a mass recreation area. And about a radio commentator by the name of Arthur Godfrey who came out to Madison to give the State Legislature "what for" because of a lot of foot-dragging on conservation action.

So you can find the Christmas spirit if you look for it.

Sammy Versus Snowmobiles

If you should hear one of these Sundays a strange crashing and roaring in your home woodlot, it will be a sign that the blissful quiet of your winter woods has been invaded once and for all by a new contrivance called a snowmobile. As you know, Sylvia, up until now a snowbound woods has been the last stand of solitude. But no more. The stretch of countryside that has lain remote and still in February is now a racetrack, and the ringing silence of far places is going the way of the passenger pigeon.

In case you haven't been reading the papers, the snowmobile is a sort of sled or toboggan powered by about a 14-horse motor equipped with an exhaust that gives off the typical stuttering roar of a two-cycle engine. The snowmobile is also equipped with one or two human passengers who get their kicks out of charging over frozen hill and dale at speeds up to 75 miles an hour. They apparently represent the strain of people who, having accomplished the conquest of our summer waters with their high-performance outboards, are now bent on subduing our winter woodlands with high octane.

Forty years ago the snowmobile was a utilitarian vehicle devoted to transporting Canadian mailmen and Antarctic explorers. Today snowmobiling is probably our fastest growing sport, complete with courses for amateurs and derbies for professionals. Wisconsin enjoys the dubious distinction of being the world snowmobile center, over 15 percent of all international sales occurring in this state alone. The appeal of snowmobiling is apparently quite insidious. Madison doctors and lawyers have formed leagues, and even the sports editor of the *Wisconsin State Journal* has succumbed, now that the Packer season is over.

The snowmobile has brought some advantages, to be sure. Game managers and wardens use them in their work, northern resorts that formerly stood idle for six months have taken on a new lease on life in winter, and people who would otherwise vegetate in front of TV sets are getting windburned instead. But in the wake of the snowmobile army is also a trail of ripped fences, trampled pine plantations, harassed wildlife, and traffic hazards. You have to express a certain sympathy for the Vilas County farmer who is currently under indictment for wounding a trespassing snowmobiler in the heel with his .22.

In the long run, perhaps, the principal victim of the snowmobile craze will be nothing more nor less than silence itself. Those of us who have taken refuge in the winter woods from the roar of the crowd have no place left. You cannot even stuff silence and preserve it behind glass in a museum. We have in Washington now a Bureau of Outdoor Recreation charged with cultivating the popularity of the snowmobile. Maybe what we need is a Bureau of Solitude.

12. Headlines and Sidelines

Dear Sylvia:

When I came down here to Madison ten years ago as the lobbyist for the Southern Wisconsin Association of Fur, Fin, and Feathers (SWAFF), it was pretty hard for me to generate any press coverage for the cause of conservation. Oh, there were a few outdoor writers who recorded what I was saying. And some little old ladies in tennis shoes would write sentimental letters to editors. But that was about all. Now times have changed, and then some. Indeed, both the scope and the velocity of the change are a little staggering. Today what we used to call conservation is big news.

Just as it happens, the term conservation isn't used so much any more. The new magic words are "environmental management" and "ecological conscience." But the message is about the same: that an expanding population, wielding a rampant technology, is wreaking havoc with our fragile natural resources. You can't pick up a paper nowadays without reading about environmental issues. What used to be the exclusive domain of a few so-called "nature" magazines is now top news in all the media. For example: *Time* magazine has a new regular section labeled "The Environment." The *New York Times* maintains a stable of environmental reporters, headed by John Oakes. *Look* magazine's recent issue is a stunning examination of America the un-beautiful. Big-time television has discovered the environment, too, devoting documentary specials to the subject.

Out at the University of Wisconsin some of my friends have even developed a new "conservation communications" major to train more and better environmental information specialists. What all this new-found emphasis means isn't quite clear at the moment. It may be the natural reaction of editors against old-hat news of moon-shots and sitskriegs. Or it may represent a long over-due recognition on the part of journalists that the biggest story of our time is the question of survival—of squirrels and of men.

Yours truly,
Sammy Squirrel

Sammy Takes a Look At SDS

SDS has discovered pollution, and the world may never be the same again.

SDS stands for Students for a Democratic Society, which may be something of a misnomer in a number of regards. Be that as it may, SDS is the heart of the radical, new-left movement on the campus. Around it orbit all sorts of disenchanted young people of various persuasions and bents. And they have been pretty successful in shaking up the Establishment. They have fired university deans, altered US foreign policy, kept burning the fires of civil rights reform, thrown a Democratic convention into a shambles, boycotted California grapes, and helped depose a US president.

Looking for new worlds to conquer, the SDS and its allies are now beginning to turn their guns on pollution. And if I were a manufacturer of pesticides or paper pulp, I'd start looking over my shoulder. At recent hearings in Madison on the merits and demerits of DDT, the SDS staged a demonstration and passed out leaflets in the classic style of today's activists. It may be only a question of time before they boycott grocery stores selling vegetables sprayed with pesticides, harass paper mills that insist on dumping sulphite wastes into public rivers, and picket a county board that dallies with the problem of Lake Mendota.

I attended an SDS meeting the other night at which such tactics were discussed. I felt right at home. The discussion sounded exactly like a bunch of deer hunters complaining about the Conservation Division. As a matter of fact, the students looked like deer hunters half way through the season—rustic shirts, beards, and all. It will be interesting to see how the old-line conservation organizations handle this intrusion in their ranks. There is much to irritate and disturb the older generation in the posture of the new left. On the other hand, in their commitment to environmental values, the SDS'ers may well represent what the apostles of conservation have been calling for all along.

John D. Rockefeller III has something to say about all this in a recent issue of *Saturday Review:* "There is a unique opportunity before us to bring together our age, experience, money, and organization with the energy, idealism, and social consciousness of the young. . . . We badly need their ability and fervor."

The ferment of youth can distract the conservation movement or it can be of enormous benefit. Which way we go will depend on whether younger people keep talking rationally—and whether older people really start listening.

Sammy Starts Protesting

I wandered into the Student Union down on the University campus the other day and beheld evidence of what may be one of the most significant developments in the history of the American conservation movement. In the corridor near the Union cafeteria down there, groups of students are always seated behind one kind of a table or another, passing out political tracts and campaign materials of various kinds. In the past, most of these student groups have represented such activities as "End the War," "Strike Against High Rent," "Join the Young Socialist League," or "Feed the Biafrans." Well, there's a new table there now, and it's manned by members of a new organization called the Ecology Students Association.

The ESA is made up largely of grad students majoring in botany or zoology, with a sprinkling of other disciplines. Its mission is to gather hard data about environmental problems, get more students excited about them, and do something. In short, the ESA'ers are smart, and they're mad. They're smart about the basic fact that man is as dependent on the health of his natural environment as are we squirrels, and they're mad about the way that environment is getting fouled up. They're mad about a campus smokestack and campus buses that foul the air. They're mad about a Navy communications project that threatens the ecological balance of northern Wisconsin. They're mad about the trans-ocean spread of DDT. They're mad about the population bomb.

Right or wrong, these new recruits to conservation are scorning the conventional conservation organizations. They're doing their own thing. And they're doing it with youthful vigor and iconoclasm. It's just a question of time before they add the tactics of confrontation to the conservation arsenal. The ESA isn't alone on the Madison campus. There's the year-old Science Students Union, an outgrowth of SDS, and there's the new Engineers and Scientists for Social Responsibility. It's all a part of a nation-wide uprising, they tell me. The new approach to conservation isn't very sedate or sophisticated, but it may just get results where gentlemanly tactics have failed. Maybe this is what Thoreau meant when he said, "In wildness is the preservation of the world."

Sammy Salutes the Sylvias

You should have been here in town the past seven days. Our human friends have been observing E (for Earth) Week. There were more speeches on conservation per hour than there've been per year since 1908. It's all a part of a new-found "ecological awareness," as they call it, which recognizes that if man doesn't make peace with nature, nature will carve up civilization into little pieces and blow it away on the winds of over-population and pollution.

The big question now is: Will the momentum of E-Week oratory get translated into long-term environmental action where it counts—in the nitty-gritty decisions of political, social, and economic life? In that regard there was one particular bright spot in Madison E-Week activities—the production and distribution of a folder, called "Household Action," which tells the average family exactly what it can do—in little ways and big ways—to stop damaging our life-support system. The folder was researched

and written by a cadre of young women, and distributed door-
to-door by the Girl Scouts. So you females deserve a special "E"
badge for "Environmental Effort."

This handy little guide tells housewives, for example, what
brands of detergents are high in the phosphates that are speeding the contamination of Lake Mendota. "Do you want sparkling
dishes or sparkling waters?", it asks. It suggests ways to cut down
on air pollution, like using an old-fashioned lawn mower instead
of a power type: "Burn fat, not gas." It has some strong words to
say against pesticides. It talks about the problems of food addi-
tives, and of solid wastes—and what to do about them in the
home. And so on. It takes only about ten minutes to read this
folder, but daily reference to its information can lead to a 180-
degree turn in your understanding of what conservation really
calls for.

In short, if anybody wants to know how he can change his
life style to conform to an ecological conscience, the "Household
Action" folder provides a roadmap and a compass. The Johnson
Foundation footed the bill.

The folder is something of a "first" for Madison. Calls for
sample copies are coming in from all over the country, and it's
being reprinted for passout in 17 towns in southern Wisconsin.
If you don't get one delivered to your bur oak, Sylvia, you can
call the E-Day Office on the University campus. Incidentally,
you know one of the folder people. She's been out to our woodlot
a number of times. Remember the girl who found a fourleaf
clover under that shagbark hickory?

Sammy Reads Some History

I was doing a bit of history reading the other day about the
food-packing industry at the turn of the century, and it all had
a familiar ring, what with another agricultural industry, the
chemical pesticide business, standing arrayed at the bar of public
opinion here in Madison. A book by Upton Sinclair, *The Jungle*,
appeared in 1906. It was a muckraking report on the food-packing business, particularly the meat-packing industry in Chicago,

complete with gruesome details about such things as hogs dead of cholera going out to the world as Durham's Pure Leaf Lard.

The book created a sensation. The public was swept by a conviction that the canned goods and other meats it was asked to buy were prepared among filth and degradation. Even President Theodore Roosevelt testified that he would just as soon have eaten his old hat as the "embalmed beef" shipped to American soldiers in Cuba. The reaction of the meat-packing industry was to deny all. It announced that *The Jungle* was "the product of a disordered and sensation-seeking mind." It persuaded the U.S. Department of Agriculture to issue a whitewash report. In a series of ghost-written articles in a national magazine, J. Ogden Armour described any federal control as "creeping socialism." But public pressure was compelling. A Beef Inspection Act and a Pure Food and Drug Act became law. *The Jungle* had achieved a permanent and constructive reform in an industry that touched and affected every human being. Today the minimum self-imposed operating standards in the meat-packing industry far exceed the first halting public steps at quality control that the industry so violently opposed 60 years ago.

My guess is, Sylvia, the time will come when the chemical pesticide industry is subject to the same sort of regulations that the food-packing business is now. At the moment the chemical people face charges before the Wisconsin Board of Natural Resources that DDT in particular is polluting the waters of the state. And they face a new bill in the State Legislature banning completely the use of DDT. The reason, in the words of Professor Grant Cottam: "Pesticides are biocides. They act on fundamental metabolic systems which are common to many living things; and when they are introduced into an environment there is no way of being certain what, in addition to pests, they will poison."

The answer of the chemical industry presently is an echo of J. Odgen Armour: "Our critics are neurotic, driven by mystic, primitive fears to the extent that they see greater reality in what they imagine than in fact." But if the lessons of history are any clue, time is running out on the indiscriminate application of toxic chemicals. The public is simply going to demand controls on pest control.

Sammy Applauds the Establishment

According to some of its youthful critics, Sylvia, the American political establishment is so moribund it cannot react quickly enough to meet modern social needs. Well, in at least one phase of Wisconsin life we have recently seen rather dramatic evidence that the establishment *is* responsive, and rapidly at that.

It was only eight months ago that the persistent pesticide DDT stood arraigned as a disastrous pollutant before a special Department of Natural Resources hearing examiner here in Madison. At the time it did not seem possible that a little coterie of scientists and conservationists could possibly prevail against the big guns of organized agriculture. As a matter of fact, a decision has yet to be rendered by the hearing examiner in question. But in the meantime the case against DDT has been taken up by everybody from Arthur Godfrey and Eddie Albert to the President's scientific advisor, Mr. DuBridge. And within recent weeks a new Wisconsin Pesticide Review Board has for all practical purposes banned the use of DDT in the state. So, when public opinion is properly marshalled, it is possible to make prompt progress in the fight for environmental quality.

It probably wouldn't have happened in the case of DDT, however, without the remarkable dedication of a little group of men and women armed with hard facts and cold nerves. There was the inspiration of Rachel Carson's book, *Silent Spring*. There were Attorney Victor Yannaconne and Professor Charles Wurster who came into the state from the Environmental Defense Fund. There were such UW professors as Orrie Loucks, Bill Reeder, and Joe Hickey, who stood up and were counted when the going was rough. There were such sources of funds as the Fishing Tackle Manufacturers Institute. And there was a bevy of coeds and Madison housewives who performed all manner of logistical chores sans pay or prominence.

It's a great story, Sylvia, and one that fortunately is being documented by a UW grad student, Bruce Ingersoll, for his master's thesis in conservation communications. The title might well be, "Good guys don't always finish last."

Sammy Writes a Speech

There's one business I wish I owned, Sylvia, and that's the business of writing speeches for chamber of commerce secretaries. A chamber of commerce is a sort of a flock or covey of businessmen and industrialists, and every chamber has a secretary, or leader, who makes a lot of speeches defending the chamber from all manner of threats. The nice thing about being in the business of ghost-writing for a chamber of commerce would be that you'd never really have to change the speech. Oh, you'd have to modify the introduction to fit the particular circumstance, but the main body of the speech you could warm over again and again.

For example, here in Madison a member of the city council has recently introduced an ordinance aimed at doing away with non-returnable bottles, cans, and cartons. The humans who live in Madison are about to be inundated in their own waste, and Councilwoman Ashman is trying to do something about it by getting rid of the throw-away containers that make up a big part of Madison's garbage. Well, they held a public hearing on her proposal the other day, and you-know-who trotted out "the speech." It goes something like this: You can't pass this or do that because it will (1) force businesses out of business, (2) make items unobtainable, (3) increase prices, (4) discourage economic development, and (5) hurt people on fixed incomes.

I think this speech was first written in 1902 to oppose workmen's compensation, and it has been used against social security and against war and against peace and so on ever since. In other words, every time anybody comes up with an idea, you can count on "the speech" to knock it down. On second thought, though, there is one part of "the speech" that does change. If the proposal is for a local ordinance, "the speech" says that the problem is a national problem and can only be tackled at the federal level. But if the proposal is for a federal law, "the speech" says that the problem can only be handled at the local level.

After the hearing on her ordinance, Coucilwoman Ashman said "the attitude of the businessmen at the meeting makes me pessimistic." In other words, she is not sure the new drive for environmental quality and conservation can carry the day against "the speech." I'm not so sure she's right, Sylvia. Once upon a time they said humans would never have fair employment practices and all sorts of other things that are "bad for business." But human society has made strides, anyway. I'm more inclined to think it's "the speech" that is just about over the hill. At least so long as there are courageous women around like Alicia Ashman.

Sammy Congratulates the IWLA

It doesn't seem possible, but the famous Boundary Waters Canoe Area along Minnesota's Canadian border is under threat of destruction again. The BWCA is a million acres of wilderness preserve in the Superior National Forest. It is the home of naturalist-author Sigurd Olson. It gained fame in the late 1940s when President Truman signed a unique order barring intrusion by air. It annually attracts hundreds of amateur Muirs and Thoreaus to its miles of motorboat-free waters. Since its establishment the BWCA has been one of our proudest examples of dedicated real-estate. As we lost other areas, we could always say, "Well, we still have the BWCA."

But now a cabal of New York businessmen propose to exercise their mineral rights and invade the BWCA to prospect for nickel and copper. There is no question that George W. St. Claire and his colleagues own ancient mineral rights in the area. The Federal Government neglected to pick them up when it bought the preserve.

But the friends of the BWCA have two things going for them: a regulation against motorized equipment, and a rule against permanent or even semi-permanent camps. Taking the lead in the defense of the BWCA is the IWLA. This may sound like a very subversive organization of miners or farmers out of the nineteenth century. But it is actually the Izaak Walton League, whose current national president is, conveniently, a Minneapolis lawyer. The League has sought a permanent injunction in the federal courts against mineral exploration in the pristine canoe area.

It is in the great tradition of the League thus to spearhead legal defenses against environmental degradation. It was the IWLA that fought the battle of Horicon Marsh, that sponsored the celebrated Wisconsin Supreme Court decision against dams on wild rivers, and that helped fund the recent attack on DDT. You simply have to ask, Sylvia, where would we be in conservation today were it not for voluntary organizations that serve as district attorneys for the rights of the public—and squirrels?

Sammy Goes to Washington

I took some time off recently, Sylvia, to go to Washington to visit our good Wisconsin friend, Senator Gaylord Nelson. It wasn't so long ago that Gaylord was practically a lone voice for conservation crying in the Washington wilderness. No more. Conservation—or, to use the new magic word, "environment,"— is now just about the hottest topic in the national capitol. In fact, the political jockeying that's going on to get aboard the environmental bandwagon is something to behold.

Not so long ago, for instance, Senator Nelson, a Democrat, introduced a novel bill that would pump federal funds into environmental education programs in the schools. Not to be out- done, a Republican Senator from New York by the name of Goodell followed up with a somewhat similar bill. Senator Nelson also took the lead in sponsoring a national environmental teach- in on college campuses April 22nd. Not to be up-staged, Secre- tary of Interior Hickel ran a series of campus seminars during Christmas vacation.

The word has it that President Nixon will focus on environ- mental problems and programs. So various federal bureaus are elbowing to assume leadership. Quite a contest is shaping up, for example, between Interior, which controls clean-water pro- grams, and HEW, which controls clean-air and clean-food pro- grams.

In the U.S. Office of Education, where conservation has been recognized for years only by a half-time lady, there is now a suite of whiz kids devoted to environmental education planning. The National Park Service is developing a strong liaison with the

National Education Association, and the Bureau of Sport Fisheries and Wildlife has recently beefed up its refuge public use program under Bill Colpitts. Even the Corps of Engineers is thinking about consulting the public more on its dam-building projects.

The one thought that bids me pause, Sylvia, is this: Will there soon be any room left for those of us squirrels who have been on the conservation train for a long time? We may just have to hang on by our toe-nails!

The Congress Rides Again

Spring is the season of the year when small numbers of Wisconsin sportsmen gather at assorted county seats to debate fish and game laws and elect representatives to Wisconsin's Conservation Congress. Although neither side would be flattered by the comparison, it is really surprising how much alike are these Conservation Congressmen and the members of student activist groups on college campuses.

While the Congress was set up over 30 years ago, it is a prime example of what today we call "participatory democracy." That is, it is sort of a people's court designed to short-stop unilateral activities on the part of either the Legislature or the Department of Natural Resources. By statute, neither the Legislature nor DNR can manage fish and game without submitting the rules to Congressional plebiscite. Theoretically, the Congress is representative of Wisconsin sportsmen. Actually, it is representative only of the handful of sportsmen who attend county meetings. Its leaders are often those extra-activists who can talk louder or sit longer than their fellows. Congressmen typically dress in "campus" attire, like plaid shirts and corduroy jeans. They come to the meetings clean-shaven, but only because they have just cut off the beards they grew while deer-hunting. Like student groups, the Congress frequently comes up with a list of "non-negotiable" demands. It has even been known to stage a protest march on the capitol, led by a real bulldozer.

Unlike student groups, however, the Congress is not really concerned with conservation issues across the board. It usually

spends its time arguing about the details of solutions to yesterday's problems. For example, you wouldn't hear much at this
month's meetings about the big population-pollution syndrome.
You won't find any women at Congress meetings either, while
on campus the activist groups have their share of females.

In the 1930s, when the Congress was born, it was a swinging
thing. Now it is "institutionalized," as they say; run, in effect as
a captive audience by DNR. It will be interesting to see whether
today's campus activist groups like the Ecology Students Association become tomorrow's claque or whether they retain their
rebel character.

Sammy Studies Economics

Mark Saturday, October 2nd, on your calendar with a big red
skull and crossbones, because that's the date the squirrel season
opens in southern Wisconsin west of Highway 78. You might
tell our grouse friends they become fair game the same morning.

I'll be safe here in Madison, because they don't permit the
discharge of firearms in the city, but your Iowa County woodlot
is apt to be a pretty dangerous spot.

If it's any satisfaction to you, no human being can go hunting
until he has paid $4 for a license. He doesn't have to take any
test, just pay his money. That's not really very much dough
these days, and yet the average hunter expects a whole lot for
his four bucks.

For one thing, he expects to be able to go hunting as many
days out of the season as he can get away from his office. Second, he expects to tramp unmolested over any hill and dale,
no matter who owns the land. Third, he expects to be able to
bring something home to eat. And fourth, he expects to be able
to tell off the Conservation Department when anything goes
wrong. All that for $4 a year.

In contrast, the very same individual thinks nothing of paying
at least $1 each and every time he plays a round of golf; he
expects to wait his turn at a crowded first tee, he doesn't expect
to break par, and he doesn't write a letter to his legislator if the
greens are rough.

It all makes you wonder just who invented the human laws of economics. The economic approach to non-residents of Wisconsin is particularly peculiar. Let's take the non-resident fisherman, for example.

First of all, Wisconsin humans spend over $400,000 tax dollars a year to lure non-resident fishermen into the state, not counting the millions of commercial dollars spent for the same purpose.

Then this non-resident fisherman is charged only $5 for a fishing license, unless he is under 16, in which case he fishes free. If a husband and wife guarantee to stay only 15 days, the two of them pay a mere $6.

Now let's say this non-resident is not a fisherman but a University student, who proposes to stay a whole year in the state, spending money like a sailor. You'd think Wisconsin humans would welcome him with open arms. Not so. In the first place, he has to be extra-intelligent before he's allowed to come. In the second place, he's charged over $1,000 for nine months of study at Madison, and another $300 or so if he spends the summer here.

Human agricultural economics are crazy, too. The farmer who owns our woodlot can get his taxes reduced by one government bureau if he fences cows out of his woodlot, and he can get money from another government agency to help him turn his marsh into a cow pasture. One government bureau will help him pay for fertilizer to grow more alfalfa, and another will pay him for not growing corn.

Say, how'd I get off on this subject, anyway? All I wanted to tell you was to be alert on October 2nd. On second thought, maybe it's about time we squirrels started shooting back.

Wisconsin Sideroads to Somewhere

As Walter Winchell used to say, "There's big news tonight!"

We're in a book! Yes, you and I are the subjects of a chapter called "Squirrels Keep Going Strong." My name is right there in big type—Sammy Squirrel. The story is all about how "more hunters probably cut their eye teeth on squirrels than on any other game." Imagine that.

The book itself is called *Wisconsin Sideroads to Somewhere*. It's written by Clay Schoenfeld, and published by Dembar Educational Research Services, Inc., of Madison. According to the author, the book is about "those adventures in outdoor recreation and conservation that are everybody's for the finding down the sideroads of Wisconsin." In the foreword, Professor Robert Ellarson says the book is "a way to enjoy the experiences that instill a respect and love for the outofdoors . . . and learn about real and vital conservation issues."

I don't know whether they've actually read it or not, but a lot of people have been saying very nice things about this book. August Derleth, author, publisher, and critic, calls it "sane, balanced, and an abiding pleasure to read." Tom Guyant, Milwaukee outdoor writer, says the author writes "with a common touch that is witty, humorous, and vivid while being informative as well."

Howie Mead, publisher of *Wisconsin Tales and Trails,* recommends the book for "the many hours of reading pleasure it will give to every Wisconsin outdoor enthusiast." The Madison Audubon Society newsletter talks about its "lots of good tips on getting more enjoyment from the outdoors, and lots of good sense in conservation."

"There is a profound message between the lines," says the *National Wildlife Federation News,* "that the development of our recreational resources is not only a matter of building wilderness preserves but also of building in human hearts a receptivity to wonders close at hand."

The Monroe *Evening Times* editor wrote that "we don't ordinarily drop a free commercial into our editorial page," and then devoted half a column to *Sideroads to Somewhere.* Both Governor Warren Knowles and Senator Gaylord Nelson have expressed a "well done," so I guess the book is non-partisan.

Author royalties from sales of our book have been assigned to college and university scholarships for budding conservation journalists through the Gordon MacQuarrie Foundation. I'll try to get you a free copy of the book, Sylvia. I know the author.

Cabins, Conservation, and Fun

Hold onto your hat! We're in another book!

Yes, we're one of the outdoor characters featured in a book by the writer of this one. It's called *Cabins, Conservation, and Fun,* published by A. S. Barnes and Company.

According to the dust jacket, the book is all about "the pleasures and pitfalls encountered in finding and developing a rural retreat." The author has drawn on personal experience and research to present a down-to-earth, step-by-step handbook for picking a cabin site, building a new cottage or restoring an old house, and conserving and enjoying surrounding resources.

It seems like half the humans I talk to these days have either already taken to the hills, are actively looking for a little place in the country, or at least are thinking about getting away from it all. It wasn't long ago that the American dream called for two cars in every garage. Today it looks like they're striving for two garages for every car. The trouble is, many folks don't know the first or last thing about finding and developing rural property. The author has tried to remedy this situation by putting down all those things that every countryman should know —things like getting to know rural real estate, checking the angles before you buy, the special problems of riparian property, how to practice conservation measures, and so on.

We're in the book in the section on "Rural Neighbors," along with chickadees, farmers, possums, and partridge. "More hunters probably have cut their eyeteeth on squirrels than on any other game," the author says. I don't know whether that's a comforting thought or not.

Just why so many people hanker to share a woodlot with us squirrels is rather curious. It may well be that for all their talk of Growth and Expansion and Bigness, a lot of humans have a certain instinct for a slowing down rather than speeding up, and a feel for the small and simple things of life. Certainly one of the country's most striking fads sees city people flocking to the out-of-doors to build weekend places. It is becoming more and more likely that 40 or 50 years from now they may have a President who was actually born in a log cabin! For these people, *Cabins, Conservation, and Fun* is "a fascinating and worthwhile adventure into the out-of-doors that gives an added dimension to leisure-time living in the country." At least that's what the publisher says.

Sammy Volunteers for Duty

There's bad news from Washington these days. It looks like wildlife conservation could become a victim of the Vietnam situation. As the Wildlife Management Institute puts it:

"World commitments of the United States and the obvious intensification of military efforts create grave concern about the conduct of natural resources programs here at home. Essential water, soil, forestry, and wildlife programs may be cut to free money for the military."

What the high-level planners in Washington may not realize is that we fur, fin, and feather folks can be very helpful to the cause of national defense. A lot of us are ready to volunteer for the duration.

Wild pigs were used in World War II, for example, as a means of clearing mine fields for troop passage. It might help if our troops had boars to test out the footing down those Vietnam trails.

Ducks and geese have been used for sentries about as long as they've been around. Gaggles of geese made infiltration all but impossible in Central and Western European villages in two wars.

Monkeys have been used for air raid wardens. They could hear and see planes before men could. If they took off jabbering into top tree branches instead of calmly eating, it was time to hunt a hole and elevate the ack-ack guns.

Hannibal's scouts once found the enemy in a deep valley, with campfires glowing and sentries out. So the wily Carthaginians got 200 long-horned cattle together on the brow of a hill, tied torches on their horns, and stampeded them downhill into the enemy camp.

The Sybarites had trained their cavalry horses to dance to a specific tune—much like some of the training given the famous Lippizaners in Vienna today. The enemy took advantage of the trait. They enticed the Sybarite horsemen to mount up, and then they played the right tune. The cavalry charge was a flop.

Dogs have been sentries for centuries. They are also frequently used on patrols.

In the days before electronic communications, pigeons carried messages. They could do it again. They have a great tradition to live up to. One fabled French bird was awarded the Croix de Guerre and the equivalent of two Purple Hearts.

You see how it is? We wildlifers are ready to go, and we won't burn our draft cards, either. And if I wind up in Saigon, Sylvia, I know you won't send me a "Dear John" letter, or will you?

Friends of Our Native Landscape

You will be pleasantly surprised, I think, to know that we members of the Southern Wisconsin Alliance of Fur, Fin, and Feathers are not alone in our fight to save the Wisconsin outdoors.

You have heard, of course, about the Conservation Department people and other state and federal officials who are on our side. What has been very revealing to me down here is the number of rank-and-file volunteer citizens who are in there pitching day in and day out for conservation.

For example, there's J. J. Werner, of 2020 Chamberlain Ave., Madison, who is chairman of the John Muir chapter of the Sierra Club. This organization helps people explore, enjoy, and protect parks, wilderness, waters, forests, and wildlife. Their current campaign is focused on "saving Wisconsin's wild rivers before the opportunity is lost forever."

William A. McGilligan, of 512 E. Gorham St., Madison, is trying to organize a Madison chapter of Wetlands for Wildlife. This organization raises money with which to keep duck nesting and resting habitat in the United States from being "exploited and destroyed at a disastrous rate."

Up in Birge Hall on the University of Wisconsin campus, Grant Cottam and other professors are active in a movement to encourage the acquisition and preservation of outstanding natural lands.

Then there are people like Mrs. Evelyn N. Warner, of 5713 Elder Place, Madison, who carried on a letter-writing campaign to keep Paul Olson of Madison on the Conservation Commission.

Now I don't want you to get the idea, Sylvia, that all humans are behind us. For example, Mssrs. Rutnik, Gehrmann, Nuttelman, Gee, Bidwell, Rommell, Shurbert, Schwefel, and Doughty have just introduced a bill in the Legislature that would take the state back to the practice of dissipating $180,000 a year in conservation money on fox and coyote bounties.

On balance, however, it looks to me like we have at least as many friends as foes around here. And the funny thing about it is, our best friends in March are very apt to be the very same guys who shoot at us in October.

Sammy Solves the Horicon Problem

Thanks a lot for that sack of fine nuts you sent me for lobbying purposes. I've already eaten them myself, however, because I can't be caught looking like I'm buying votes, even for a good cause like conservation.

Let me tell you what's happening to our friends, the wild geese.

You remember my talking about Horicon Marsh. That's a big tract of swamp about 30 miles north of Madison in Dodge County. Years and years ago Horicon was a great natural waterfowl haven. Then for a time it was ditched and drained into a desert. Now, thanks to the efforts of many outdoor fans, Horicon has gradually been restored. The trouble is, it's too good.

If you were a goose, I guess Horicon would look like the finest motel in the country. It's about half way between Hudson's Bay and the Caribbean. It has acres and acres of cattails, fields, and water. It is surrounded by some of the finest corn plantations in the world. Something like 50,000 geese or more stop there each spring and fall.

That was the whole point of restoring Horicon—to provide a big refuge for lots of migrating Canadas and bring back a little goose shooting. Now so many geese use Horicon that the orderly management of the Mississippi flock is jeopardized.

So what is the government going to do? Well, after trying for years to bring the geese back to Horicon, it is now going to try to get them to go away. The government boys have stopped growing food crops on the refuge to help prevent a big build-up of birds, and they plan to "haze" the geese this fall to drive them south on their migration route.

After living in Madison off and on during the past few years, I have some ideas about how you can render Horicon Marsh unpalatable. For one thing, you could install parking meters. You could announce you were going to build an auditorium, and then argue for ten years where you're going to build it. Or you could dump all manner of refuse into the water until it turns pea-green. You could install stop-lights to frustrate the geese as they fly to and from the cornfields. Better yet, you could lay out a subdivision and a golf course in the middle of the marsh.

I really don't think our human friends will have much trouble getting the geese to leave Horicon once the experts put their minds to the problem. All humans have to do is act naturally, and they can blight any garden spot.

Sammy Views the Election

They had a big election this month, and they let me cast a ballot. Believe me, it took some persuading at the registration office. The man behind the desk said at first I didn't qualify because I was a squirrel, but I simply pointed out the law doesn't specifically say a squirrel is not a citizen. Then he said I wasn't old enough. So I got a professor of zoology at the University to testify that a three-year-old squirrel is the same as a 54-year-old human. That fixed me up.

Deciding who to vote for was a problem. As you can imagine, I was mostly interested in electing somebody who would be a good friend of us fur, fin, and feathers folks. Somebody told me Mr. Nixon was a "hawk" in disguise, so I decided to vote for him. But then another person told me Mr. Humphrey was a "dove" at heart. So I didn't know what to do. The party symbols didn't help. One is an elephant and the other is a donkey.

At any rate, Mr. Nixon won the presidential race. Just what

this means for conservation, it's a little hard to tell. Conservation was born under a Republican president, Teddy Roosevelt, 60 years ago, but in more recent times it has been the Democratic administrations that have evinced a concern for our natural resources. President-elect Nixon's only conservation speech was made on May 16, 1968, at Portland, Oregon. In it he said:

"Appropriations for conservation should escape the budget knife. . . . We must find the time and energy to preserve and improve our heritage. . . . The vital question is not the quantity of our resources but their quality—and their effect on our lives. . . . We need to develop objective standards of environmental quality, and effective, fair means of enforcing them."

Fortunately, one of our best friends will be back in Washington—Senator Gaylord Nelson. He based his campaign in large part on his distinguished career as a conservationist, first as governor of Wisconsin and more recently in the national capital. The people of the state responded with a big vote for Gaylord. For the last ten years Senator Nelson has been in the forefront of the fight against water pollution, air pollution, soil erosion, wetland deterioration, waning wildlife, urban sprawl, disappearing wilderness, and blighted landscapes. He has played a key role in earmarking land for outdoor recreation, in saving wild rivers, in reserving lakeshores, in protecting endangered species, in blocking indiscriminate pesticide spraying, and so on.

So long as we have representatives like Gaylord Nelson, we squirrels know we have a strong voice in Washington.

Sammy Gets Stuck in Cherokee

I'm going to attend an important meeting this week between the Wisconsin Conservation Department and the city of Madison, having to do with the future of Cherokee Marsh. Some of my friends advise me not to go; they say if I get stuck in Cherokee, I'll never get out. But I'm going anyway, because what happens to Cherokee Marsh "will have a profound effect on public policy with respect to stream and wetland areas throughout the state," as the League of Women Voters says.

Cherokee Marsh, in case you don't remember, is 3,000-plus

acres of riverbank, pasture, woods, highlands, meadows, fields, swamp, thickets, muck, lake, and cattails, lying just to the northeast of Madison where the Yahara river empties into Lake Mendota.

Since 1959 Cherokee has been earmarked by local planners as a natural area to be retained in public ownership for outdoor recreation and conservation. Public control of the marsh is particularly important as a means of protecting Lake Mendota from additional pollution and siltation.

Well, a couple of years ago city people, federal people, and state Conservation people got together and entered into a compact to buy Cherokee Marsh and turn it into a multiple-use preserve and park at Madison's doorstep. The bill was going to come to something like a million and a half dollars.

Since that time two things have happened. First, in a complicated deal with the city council, a group of local entrepreneurs known as Cherokee Park, Inc., have won permission to proceed with residential and commercial development on the part of Cherokee lying closest to the city limits. Second, land prices in the whole Cherokee area have skyrocketed.

Disappointed but undeterred, the city has been proceeding with its part of the compact, buying and condemning the Cherokee lands it had agreed to acquire. The Conservation Commission, however, has been dragging its feet.

That's the purpose of the meeting this week—to try to find the reason for the "delays and indecision" on the part of the Conservation Department, as Madison Assemblyman Norman Anderson describes the situation. He predicts "a loss of incalculable public esteem and support" were the state Conservation people to pull out of the Cherokee project.

I guess you could say the position of the Conservation Department in Cherokee is sort of like the position of the United States in Vietnam. They made certain promises, and now their honor is at stake. Our Southern Wisconsin Association of Fur, Fin, and Feathers hopes they come through.

Sammy Wonders About USDA

The Legislature has finally gone home for the year, Sylvia, so I have plenty of time here to look around, read the papers, and talk to people. And what I see and hear isn't very good. Our human friends are in a bad way. So many Americans—something like three-quarters—have crowded into two percent of the landscape called cities that air pollution, water shortages, highway congestion, slums, slurbs, and crime are just about grinding things to a halt.

When too many of us squirrels wind up in the same woodlot—when we, as they say, exceed the carrying capacity of our habitat—some pretty disastrous things happen to us, like predation and starvation. But human beings have something called conscience or ethics that prevent them from letting natural laws operate, if they can help it. So they are looking around pretty desperately these days for solutions short of letting nature take its course.

One idea, being promoted by Secretary of Agriculture Hardin, goes like this: Rural America is losing population. There's all that open space and there are all those small towns going to waste, while people crowd into cities because of the services and the arts and the amenities that cities can offer. To reverse this trend, to hold people in the countryside and to promote migration from big cities to small towns, why don't we, says Mr. Hardin, take the city to the people instead of bringing more people to the city. In other words, Sylvia, he's out promoting turning the Mount Horebs and Barnevelds and Arenas of the country into miniature Detroits. In cooperation with the counties and the state universities, he is placing so-called "development agents" in key spots, and he is making a lot of money available for subsidizing the transplantation of industry from the inner cores to the crossroads.

In all this planning, Sylvia, I wonder if anybody has thought about us squirrels? After all, we're part of a landscape that makes the countryside attractive, but if we have to compete with super-highways and smokestacks, I wonder how much longer we're going to be able to hang on. It wasn't so long ago that our big enemy was the farmer who hacked down den trees and pastured his woodlot. Now our big enemy is the developer and his bulldozer. Interestingly enough, the federal bureau back of both of them seems to be the United States Department of Agriculture. Wouldn't it be better if our human friends spent more money on salvaging the cities and on conserving the countryside, rather than on trying to turn the whole landscape into one gray pallor of "progress"?

Let's hope the President's new Council of Ecological Advisors can get through to the USDA.

Sammy Appoints a New Committee

You may think there isn't much doing down here at Madison when the Legislature is not in session but actually this is a very busy time of the year, thanks to committee meetings and conferences all over the place.

Aldo Leopold once complained that conservation consists mostly of "letterhead pieties and convention oratory." He may have been right. At least I find myself attending one heck of a lot of hearings, all of them supposedly having something to do with conservation, many of them producing little more than printed proceedings.

You even need a score-card to keep track of the players in this business, because there are all sorts of groups and causes and people, all of them devoted, they say, to the "wise use" of our natural resources.

Some of these people are "resource-oriented." That is, with them the husbanding of natural amenities comes first. Other advocates are "people-oriented," and talk about the greatest good for the greatest number. Still others are "money-oriented," and say that what's good for business is good for the country.

At one end of the spectrum, for example, are the hard-line "conservation" boys, who tend to emphasize protection and preservation of fish and wildlife. At the opposite end of the scale are the exponents of "resource development," who tend to look for maximum returns from tourism. In the middle are people like the "outdoor recreationists" and the "regional planners," who are concerned with "esthetic appreciation," "environmental design," and things like that.

I guess all together everybody in the conservation movement is really on the side of the angels, but it gets pretty confusing to a squirrel who's just trying to save his woodlot from a cross-cut saw.

All of these people do a lot of talking, but not always much communicating. New ideas and programs do really come into being here, but only a computer on very friendly terms with all the layers of decision-making in Madison could tell you exactly how ideas become actions. There's even much talk about talk itself.

A conference chairman the other day, for instance, opened the meeting by quoting the crusty farm leader of the 1930s, George Peek. Peek once said that "the common characteristic of all uplifters is an unquenchable thirst for conversation; they are all chain talkers." Whereupon the chairman gave a talk.

Still, as the Bible says, in the beginning was the word. The problem is to make sure that the ending around here is not also the word. Sometimes that seems a clear and present danger.

The amount of cross-fertilization—or down-field blocking—of ideas and new facts, in speeches, reports, seminars, and press releases from the institutes, agencies, centers, and committees, which breed as fast as wire coathangers, has now become a problem for space and fire prevention experts.

As the groaning condition of my desk, and all other desks in this capital, testifies daily, we obviously need a new committee to screen other committees.

Part V
EVERYBODY'S ENVIRONMENT

13. The Third American Revolution

WE ARE EXPERIENCING WHAT YOU MIGHT CALL THE THIRD AMERI-
can revolution. The first American revolution 200 years ago was
of course a political revolution; the second American revolution
a hundred years ago was a technological revolution. The third
American revolution is a cultural revolution, characterized by a
new man-land ethic. The spirit of the '70s is a spirited concern
for environmental quality. We are figuratively and literally sick
and tired of a mis-development of America that diminishes daily
the quality of the human experience: water pollution; air pollution; soil erosion; forest, range, and wetland deterioration; waning wildlife; urban sprawl; preempted open spaces; vanishing
wilderness; landscapes scarred by highways, litter, noise, and
blight—a not-so-quiet crisis of decreasing beauty and increasing
contamination that threatens not only the pursuit of happiness
but life itself. And we are beginning to *do* something about
environmental quality conservation, redevelopment, and maintenance. How we got into this revolution and where we are
going is the story of one of the crucial currents in American
history.

Environmental Attitudes

At least five gross attitudes can be identified in American man-
land relationships. The first is the *economic* impulse, which tends
to view natural resources essentially as a God-given stockpile to

277

be manipulated largely for private profit. The second is the *evangelical* attitude, which prefers to sense in nature a mystic shrine that is to be adored but not altered. The third attitude is the *aesthetic-athletic*, seeking enjoyment of one kind or another in the out-of-doors. The fourth is the *apocalyptic* attitude, which sees a mushrooming population, a rampant technology, and a fragile biosphere on a collision course toward inexorable catastrophe. The fifth posture is the *ecological*, under which we recognize the validity of each of the other approaches and attempt—through some sort of scientific understanding, technical design, and social entente—to achieve a viable balance that will protect without penalizing, develop without destroying.

These attitudes do not necessarily exist in a pure and exclusive state either within a group or within an individual. A farmer, for example, may take an economic approach to his fencerows and an aesthetic-athletic approach to the squirrels in his woodlot. A community not infrequently assumes an economic attitude toward a nearby river and an evangelical attitude toward the ladyslippers in an adjoining arboretum. A state government may adopt an ecological attitude toward resource conservation and development planning, and an economic posture toward actual resource use. That the five approaches are thus at one and the same time competitive and complementary renders man-land relationships frustrating for the biologist, fascinating for the social scientist, and functional for the politician.

While all five approaches to man-land relationships have existed for many years and exist side-by-side today, each has had an ascendancy at a particular time. Prior to the turn of the century in America, the economic motive was dominant in resource use. With an almost religious fervor, all of society acquiesced in an *exploitation* of natural resources. As William Freeman Vilas expressed the philosophy of the day, if God had not meant Wisconsin's virgin pineries to be clear-cut to build Chicago, He would not have caused the major rivers of the state to flow southward. The government bestowed various bounties on the more adroit explorers; society reserved its accolades for the barons of rampant industry; the universities devoted some of their growing skills toward a more efficient assault on field, forest, and mine.

While the exploitation of natural resources can scarcely be said to have been halted, since about 1900 the sheer economic attitude toward man-land relationships has had to answer increasingly to the evangelists and their vision of nature's "vast, pulsing harmony." The evangelical attitude has found its expression in three principal developments: the scientific study of the dynamics of nature; the *preservation* of such natural phenomena as rare scenery and remnant wildlife populations; and a populist revolt against monopolistic practices of resource waste. Confronted with these thrusts, the economic attitude became tempered by a concern for resource management more in the public interest. Using as its touchstone the term *conservation,* this wedding of research and reform has seen attempts at manipulation for wider and wiser use of soil, water, forest, range, and game. State and federal bureaus have been founded to foster various brands of conservation, and the courts have rendered several classic decisions defending the integrity of the biota. Schools and colleges came to devote a portion of their enterprise to conservation education. Conservation creeds became linked to broad programs of arid land reclamation, water power development, public welfare, economic pump-priming, and national defense. Indeed, by becoming all things to all men, the conservation movement tended to lose some of its viability as a man-land ethic.

In more recent years, the aesthetic-athletic attitude toward man-land relationships came to be heard from in swelling tones. The well-nigh cabalistic word for it was *recreation*. Recreation enjoyed an "in" phase as a catch-all term to describe the manifold outdoor activities of millions of Americans, and as a politically attractive device for developing a region's economy. The term should not be dismissed as mere phrase-making. Its wellsprings are deep in the American psyche. Americans and their out-of-doors have been carrying on a love affair for a long time. It is no accident of the songwriter's art that two of our national anthems speak of "woods and templed hills," "spacious skies," "amber waves of grains," and "purpled mountain majesty." It is of some significance that the most popular movie and record album of all time start out with magnificent views and strains of "hills alive with the sound of music." Americans simply pos-

sess this deep yearning to get into the outdoors. Be he a Thoreau
or a lathe operator, when an American looks for meaning in life,
he seeks it not in ancient ruins or in the canyons of a city but in
a forest, by a river, or at the edge of a lake. The outdoors is a
source of inspiration and, literally, re-creation—a renewing experience, a refreshing relief from alarms and routine, plus a dab
of physical exercise.

Scarcely had recreation had its day as a phrase with political
sex appeal when it was overtaken at the turn of the decade by a
revival of Malthusian fear. The new Jeremiahs, with their apocalyptic warnings of impending disaster, substituted sheer human
survival as the basic motivation in man-land relations. Over-
pollution and over-population, they said, render man the most
endangered species of all.

These four attitudes toward man-land relationships—the economic, the evangelical, the aesthetic-athletic, and the apocalyptic
—have led to two principal and competing doctrines of conservation. The first, what Professor Samuel P. Hays has called the
"gospel of efficiency," has perhaps never been better expressed
than by forester Gifford Pinchot in his dictum that land and
water use should be governed by a concern for "the greatest
benefit for the greatest number of people for the longest possible
time." This principle runs hard up against a second view of conservation, what might be called the Muir-Leopold thesis, that
what we should really seek is "a state of harmony between man
and land," in the attainment of which we must recognize that
land and water, as well as people, have certain inalienable rights.

The schism inherent in these competing philosophies of conservation is rendered even more intense by the fact that modern
conservation issues seldom involve raw exploitation versus preservation, but rather excruciating conflicts between prudent use of
resources for one acceptable purpose versus prudent use for another, as a growing population impinges on a shrinking or at
best a static, resource base. Irrigation versus power, access highways versus wild rivers, logging and grazing versus down-stream
siltation, potato plantations versus trout streams, wheatfields versus duck factories, pulp mills versus clean water, artificial lakes
versus ancient canyons, camp grounds versus forest cloisters,
smokestacks versus prairie vistas, suburban sprawl versus park-

lands, air conditioned offices versus power plant pollution, automobiles versus smog, convenience packaging versus mountains of garbage, worm-free apples versus silent springs—the list is unending and overpowering.

For example, even the so-called outdoor recreation movement can hardly be considered homogeneous. On the contrary, it is composed of a wide variety of essentially competitive activities and values. The water skier, the fisherman, the duck hunter, the bird watcher, the lotus fancier, the cottager, the hermit all compete for the same patch of water. The quality of a mountain view for one hiker is in inverse ratio to the number of campers who seek to share that view. Indeed, high-density outdoor recreation has the capacity to destroy the very enjoyment it sets out to capture. The number of fans at a football stadium does not detract from, and may even enhance, the sport. The same cannot be said for grouse hunting on a back forty or even for picnicking in a park.

Importantly, the contending parties in conservation are not asking simply, "When are we going to 'run out' of my resource?" Long before we are aware of an impending "running out," we are becoming acutely aware of something else that is happening to us—a deterioration in the *quality* of the resource. The basic issue in resource conservation today is hence not quantity but quality. The great concern in the management of resources has become the maximization of quality of output. It is not the quality of the resource itself that we are concerned with so much as its capacity to enhance the quality of life. This is a very sticky problem, as economist Ayers Brinser said, and becomes involved in many subjective evaluations.

It is the making of sophisticated choices, then, the rendering of subtle value judgments, that is the essence of conservation today. Should we do it, and if we do it, what do we gain, at what cost, and what do we lose, at what cost? These are the questions, and the term "cost" is being used in the context of physical and mental health as well as simply in reference to dollars. In the words of geographer John C. Weaver: "Conservation has become more ethics than economics."

Where do we start? What azimuths shall provide our orientation? Whose standard of quality shall we adopt? for what?

when? where? How indeed do we begin to make wise choices? How do we develop without destroying? How do we protect without penalizing?

Enter the *ecological* attitude toward man-land relationships— the attempt to balance the demands of an industrialized, urbanized, lavender-plumbing society against the demands of the living landscape, both of them being basic to modern health and happiness. This attitude thinks of man *in* nature, rather than of man *and* nature. It is the long, integrated view: Landscape is something to enjoy, not merely to mine or till. Space is something to roam in, not merely to fill. Beauty is something personally to cultivate, not merely to read about. Flora and fauna are something to cherish, not merely to harvest. Water and land and air are resources with an all-too-fragile integrity that defies any doctrine of anthropocentrism. Yet man is something more than an animal, whose spirit merits something more than a niche on the African veldt.

Environmental Programs

As never before, Americans are coming to appreciate this "oneness" of the elements of their environment—that insects, birds, fish, animals, water, soil, wilderness, trees, plants, and man are all part of the same scheme of nature, a sort of intricately woven fabric; snip one thread and the whole thing begins to unravel. Americans are coming to appreciate as well a continuing and intimate relationship with their natural surroundings that surmounts the curtains of civilization. New Yorkers, for example, during the drought of recent five years, saw air conditioners silenced, lawns browned, and water glasses banished from restaurant tables, while the Hudson River was daily carrying 11 billion gallons of undrinkable, uncleanable water past the city and dumping it into the ocean.

As seldom before, Americans are expressing a deep concern about the management of their environment. The public prints have made "the rape of the land" a headline story. Our affluence, our general values, and our social objectives are beginning to permit us to make viable choices respecting the utilization of natural resources. We no longer assume that all land and water must inevitably be devoted to the basic sustenance and protection of human life. We are ready for what Professor Philip Lewis, Jr. calls "a second integrated look" to identify the meaningful natural and cultural resources which, if protected and enhanced, can provide many types of environmental experiences for richer living, working, playing—and survival.

As seldom before, Americans are acting to conserve. Too often, the act may be too little, too late, but each act is at least some evidence of faith, hope, and maybe even love. Legislative bodies at all levels of government, public agencies, and private groups are seeking answers to the difficult questions posed by multiplying man and disappearing land. The alarm has been sounded by senators like Gaylord Nelson, calling on "the energy, idealism, and drive of the oncoming generation" to save us from "the poisonous air and deadly waters of the earth." The alarm has been sounded by ecologists like Paul Ehrlich, asking us to see "the connection between growing population and steady deterioration of the quality of life before our planet is irreversibly ruined." The alarm has been sounded also by hucksters like Arthur Godfrey intoning that "our country's highest priority in the 1970s must be survival." Epitomizing public response to such warnings is the establishment, under a 1969 National Environmental Policy Act, of a three-man Council on Environmental Quality to "advise, assist, and support the President of the United States on all environmental concerns."

To describe these ecological efforts to come to grips with the degradation of man's interlaced surroundings, the term *environmental programs* is increasingly entering the lexicon of the country. Some might say this term has sprung into being merely to lend a charismatic quality to the matter with which it is associated. On the contrary, the term is coming into use to satisfy the very real need of scholars and administrators to describe, if not a new program or set of programs, at least a new way of looking at a variety of old programs, their relationships, and their potential contributions.

While no pinpoint definition or delimitation of the term "environmental programs" is possible at this time, and indeed may never be desirable, we can list the factors or criteria that seem to be implicit in the use of the term.

First, we are concerned with the environment of *man*. It is possible, of course, to study the physical nature or the biological characteristics of the environment on an infra-human basis, but the concept in "environmental programs" is the study of man as he affects and is affected by his environment, for good or ill. The focus, in addition, is upon the growing number of humans concentrating in increasing densities and bringing greater pressures to bear upon the environment.

Second, we are concerned with the *total* environment: its social, cultural, economic, and aesthetic, as well as its physical and biological, aspects. To seek an optimum total environment requires both an understanding of human needs and the needs of a healthy living natural environment. Any discussion of the goals of society must quickly draw upon the knowledge of the nature of the world man lives in, just as any discussion of a balance of nature today must take into account the necessary impingements of man.

Third, we are concerned with *interdisciplinary* programs. The development and management of an optimum total human environment requires an understanding of the contributions that can and must be made individually and collectively by all the arts, sciences, and professions.

Fourth, we are concerned with integrated programs that have as their ultimate rationale the development of *open-ended solutions* for environmental problems, rather than short-term approaches that may actually degrade the environment. We are concerned with the adjustment of designed time and space for optimum human performance within the carrying capacities of the environment. The desired objective is to bring conflicting forces into functional relationships, resulting in a unity called order, an order where human impact does not needlessly destroy environmental quality and where environmental quality contributes to more fruitful human life, liberty, and the pursuit of happiness.

Finally, while we recognize the essential importance of

strengthening existing programs, we look toward teaching, research, and service configurations that will *transcend* traditional
lines of endeavor, and be concerned with the wholeness of the
relationship between man and the total environment. What we
seek is integrated environmental management based in the scientific method and expressing aesthetic dimensions.

Environmental Problems

What are the major problems confronting integrated environmental management today? Using the state of Wisconsin as a
parameter, major problems in environmental management today
could be defined to include the following:

Pollution will certainly show up on anybody's list of major
environmental problems. There is the problem of gross municipal
and industrial water pollution; the problem of more subtle over-
enrichment of water from agricultural and residential sources;
the problem of air, land, water, wildlife, and even human pollution from the use of chemicals as pesticides. The harmful effects
of all types of pollution must be determined through continued
research and eradicated through improved technology and control.

By any standard, another major land and water use problem
in Wisconsin today is *soil erosion.* Despite over 25 years of SCS
efforts, we have completed only about a third of the erosion control job in Wisconsin. Scarcely one-quarter of the landowners in
Wisconsin have conservation plans, and many of the existing
plans are only haphazardly followed. Soil erosion is a serious
problem on over six million acres of Wisconsin farmland. Top
soil continues to disappear, robbing us of food and fiber and
silting our streams, lakes, and wetlands.

A third major problem is *timber land management.* Some suitable acres of Wisconsin could yet be replanted or newly planted
to trees. Many unsuitable acres already have been or are being
replanted or newly planted to trees. Many acres of extensive
forest land require timber stand development measures, applied
with an eye to multiple use. The many widely scattered wood-

lots that account for some 60 percent of the state's timber acreage need protection from encroachment by unsuitable land uses,
as well as good forest management practices.

What may seem like a narrow problem yet one with wide
repercussions is the problem of *waning wetlands*. Wetlands are
those little pockets of damp countryside characterized by a high
water table and heavy emergent vegetation. They are vital to
upland game bird, furbearer, and waterfowl populations, are
relevant frequently to game fish production, play a role in the
recharge of underground aquifers, act as pollution-filters for
lakes and streams, and invariably are associated with various
forms of outdoor recreation. We have lost wetlands at an alarm-
ing rate over the past four decades to agriculture and urbaniza-
tion, in some counties as much as 90 percent. The preservation
of a viable array of wetlands is absolutely essential to a healthy
Wisconsin landscape.

A fifth problem is what we might call collectively the *crime
of the city*. It has two aspects: one, the noise, congestion, dirt,
foul air, foul water, and general lack of amenities in the inner
city—uncultured blight; and two, the encroachment of the outer
city and its appurtenances like transportation systems, dumps,
septic tanks, and "slurbs" in nearby cropland, open space, woodlots, wetlands, and recreational areas—unintelligent sprawl.

A sixth broad problem is the *degradation of outdoor recreation
areas* through improper and conflicting land-use development in
relationship to agricultural practices, parks, highways, waterways,
resort areas, natural areas, beaches, and so on; to the point where
water, fish, game, scenery, and other recreational resources are
being destroyed by the very people who seek them out, where
persistent unintelligent use of floodplains subjects people to economic loss and even loss of life, where unzoned shorelands are
taking on the appearance of outdoor slums, and where fewer
people have access to a natural heritage of decent quality.

All of the problems so far listed concern man's stake in the
man-land equation. There is a seventh collective problem in
land and water use; namely, *the preservation of the biota* for the
sake of the biota itself. Here our concern is with scientific areas,
wild rivers, remote forests, rare flora, disappearing birds. Or, if
you are unhappy with simply husbanding nature for nature's

sake, call this problem the problem of *preserving natural beauty* —what Stewart Udall described as "our groping for something we cannot forget—the long waves and the beach grass; white wings on morning air, and, in afternoon, the shadows cast by the doorways of history."

We can say some things about this list: First, it is parochial. Other states have somewhat different problems. Second, it represents a gross over-simplification of what is in reality a congeries of problems that could easily become a list of four score or more. Third, it tends to identify symptoms of disease rather than the disease itself. What *really* causes our basic problems in land and water use?

At bottom, of course, is the simple fact that we have too many people in the wrong places. If we don't practice some form of population control and redistribution, all other approaches to conservation will be for naught.

There are some other profound reasons for our current land and water use problems. Robert Ardrey suggests one of them in his book, *The Territorial Imperative*. It is seemingly a biological law that an individual or a troop stake out a piece of real estate—a territory—and call it their own. Those of us who go out into the country to buy a lakeshore lot or a back forty, to do with it what we will, may think we have some very advanced economic or cultural motives for so doing. In reality we are simply reacting to the territorial instinct that our ape ancestors acquired on the African veldt. The institution of private property, with all its attendant threats to resource management in the public interest may, therefore, be as immutable as our interior bone structure.

Our Judeo-Christian tradition complicates conservation. Unlike a heathen, who has no concept of "ownership" and who views himself as simply one of the many manifestations of nature, we have been given "dominion over the fish of the sea, and over the fowl of the air, and over the cattle, and over all the earth, and over every creeping thing that creepeth upon the earth;" and we exercise that dominion as a coach rather than as a team player.

Our economic system complicates conservation. Frugal land and water use requires us to manage our resources at least in part for other people who can't or won't pay us because they are either not direct customers or they are not yet born. There is nothing or very little in our system of interest rates to encourage such management. In fact, under our pervasive doctrine of conspicious consumption, we are driven in just the opposite direction, to make a fast buck today by catering to the demands of a heedless market.

Our tax practices often mitigate against wise land and water use. We persist in billing the wrong people too much and the right people too little for conservation services rendered by the state. We are often hamstrung in reform by constitutional provisions, irrelevant precedents, and unrepresentative legislative bodies.

Our research posture complicates conservation. For one thing, domestic research has to some extent gone out of style. Our leading scholars are increasingly devoting themselves to the theoretical concerns of the international scene rather than to the workaday problems of the state. For another thing, while our colleges and universities house a number of disciplines related to land and water use, each department has an intradisciplinary approach. What has been lacking are interdisciplinary studies concerned with the total relationship of man and his environment.

Our attitude toward teaching complicates wise land and water use. We are so objective we define no right or wrong, assign no obligations, call for no sacrifice, imply no change in philosophy of values. We have been too timid, too anxious for quick success, to tell our students the true magnitude of their obligations. We are prisoners of antiseptic concepts. We march up to the moment of decision and then turn and run.

Our action techniques complicate wise land and water use. Too often we seize on over-simplified solutions, cloak them in catch-words, and peddle them like lightning rods to customers who apply them with religious fervor and then wonder why earthly salvation continues to elude them.

Our instrumentalities of government complicate wise land and water use. State, county, and municipal boundaries were laid out at the whim of surveyors a century ago; they bear little or no relationship to the configurations of the land and water

with which units of government must deal. We desperately need new political and social devices that stand a decent chance of translating value judgments into action on the living landscape.

Listen to these words: "The true enemy of preservation of our environment is our system of government, particularly local governments and county governments that are entirely dependent upon the property tax and the payroll structure. Conservation can never be accomplished so long as local government must as a means of its financial survival get new development into its boundaries." And those are not the words of a Berkeley fanatic; they are the words of a Republican Congressman.

All of these problems contribute to one central problem: unilateral approaches to land and water management. Time and again we persist in applying to our environment a practice which, while it may be beneficial for one purpose, is deleterious for other purposes. Our governmental agencies, our educational institutions, our socio-economic patterns, our cultural standards are all seemingly in league to force us into these unilateral approaches, and hence to introduce or aggravate land and water use conflicts. Somehow we must break out of our assorted straight-jackets to take an integrated view of our surroundings and carry out programs that truly reflect the "oneness" of our environment, its problems, and its needs.

Ecological Hope

That such ecological thinking is possible the public prints bear increasing witness.

Ecologist Leonard Hall asks us to think large about "the dilemma which mankind faces today":

1. The danger of planet-wide environmental poisoning from nuclear fallout and nuclear waste disposal; or of a nuclear or biological holocaust that would end the world we have known.

2. The danger of famine, starvation, and pestilence affecting hundreds of millions of human beings in the world's unproductive areas, if we cannot bring about a drastic and revolutionary flattening of the world's population curve. And while it is easy to say the consequences cannot reach our rich and comfortable enclave,

this ignores the fact of the chaos that would sweep the planet.

3. Destruction of the environment for humans and countless other living creatures through the poisoning of soil, air and water by sewage and industrial wastes; by combustion and overuse of fertilizers, pesticides and herbicides; and by erosion and soil exhaustion caused by today's monoculture and the continued stripping of forest cover.

4. Destruction of the esthetic environment—which is to say, the *quality of life*—through sustained attack on wilderness and wild nature, on wildlife, on the beauty of the rural scene, and on the equally essential and largely neglected orderliness of the urban and suburban areas where 75 percent of our people live today.

Forester Edward C. Crafts asks us to think large about "hard unavoidable steps to reverse the trend of environmental retrogression":

Population control, higher taxes, higher consumer prices, lower corporate profits, lower material standard of living, revision of national priorities, and coercion.

Look Magazine asks us to think large about an "agenda for survival":

Man's quarrel with nature is nothing new. It is rooted in the Book of Genesis, in God's command to "be fruitful and multiply, and fill the earth and subdue it; and have domain over . . . every living thing." Armed with this injunction and a mischievous new technology of awful power, we have multiplied recklessly and asserted our dominion by random slaughter. The question now is: What can we do to repair the damage we have done and avert future disaster?

First, we must get rid of the notion that the rest of Creation exists only for man's convenience and profit and that other forms of life are somehow inferior—enemies to be conquered, harnessed, or crushed. The fact is, man is just one member of a natural and interdependent community of every living thing.

Then, we have to check population growth. Without an immediate commitment to an effective program of birth control, the underdeveloped world is doomed to death by famine, and the affluent world to social chaos.

We must take stock of our planet's resources to learn how much and what kind of development our environment can sustain and how best to protect the irreplaceable wilderness we have left. Such an undertaking will have to be coordinated internationally by the UN—and might, in fact, give that body a new lease on life. In America, we should strengthen the President's Environmental Quality Council and create like groups at every level of government.

And we must clean up the mess we have already made. Industry and government together must restore our polluted air and water and our defiled landscape. The cost of a healthy environment must become part of the basic cost of doing business.

We need not be discouraged. When threatened, man is capable of almost anything. Nothing less than our survival is at stake. The problem is getting enough people to realize this blunt truth while there is still time to act.

Eighty Congressmen ask us to think large about what we can do as individuals:

Among citizens, we turn to youth as the great hope for the Environmental Decade. Young people are understandably outraged by the cynicism and materialism of their older generation. We urge them to substitute constructive impulse for negativism, and to build for future generations an environment worthy of free men and women. We hope they will conduct studies, sponsor educational forums, initiate petitions, support court suits, pressure administrative agencies, and draft legislation, as well as do the many private things needed to help protect against environmental destruction.

All American institutions and individuals must adjust their functions and policies in the spirit of the quest for environmental quality. To paraphrase Leopold, barring love and war, few enterprises continue to be undertaken with such abandon, or by such diverse individuals, or with so paradoxical a mixture of appetite and altruism, as that group of vocations and avocations known collectively, precisely or not, as conservation. It is, by common consent, a good thing to practice conservation. But wherein lies the goodness, and what can be done to encourage its pursuit? On these questions there continues to be confusion of counsel, and only the most uncritical minds are free from

doubt. America certainly has as yet no magic formula for cap-
suling conservation and administering it to assorted people and
places. It increasingly recognizes the importance, however, of
confronting researchers and implementers with sets of resource
management principles and values, and encouraging professors
and public officials alike to face the broad environmental problems upon which the American public is being asked to render
crucial judgments. Perhaps out of this flux will come integrated
programs and techniques based on, and consistent with, a syn-
thesis of new knowledge in both the natural and social sciences,
and which will find their expression through public policies; private management decisions; actions of business, farmers, and
labor; consumer behavior in the market; and voter behavior at
the polls. In essence, we are coming to address ourselves to
laying a basis for actions, to elucidating the choices in land and
water use and relating them to general values and social objectives, to instilling in people a desire for constructive change,
and to providing practical guidelines characterized by integrated
approaches. This is the changing role of America in conservation,
as American life and American learning proceed together toward
what can yet be broader lands and fairer days.

It may well be that modern man can never achieve complete
harmony with land, any more than we shall achieve in our day
justice or liberty for all people. In these higher aspirations, as
Leopold said, the important thing is to strive. We must cease
being intimidated out of hand by the argument that an action is
impossible if it does not yield quick profits, or that an action is
necessarily to be condoned because it seems to pay. That philosophy is dead in human relations, and its funeral in land relations is long overdue. The third American revolution is the appearance of this ecological conscience.

14. The Ecology of the New Conservation

WHEN THE VERY FIRST SETTLERS CAME TO THE REGION AROUND THE tip of Lake Michigan in the early 1800s they found much of the land covered with "oak openings," or savanna—a striking combination of scattered, mature trees amid prairie patches, the whole array appearing, in the eyes of one early observer, "like so many old orchards."

The trees were principally bur oaks, their characteristic thick bark protecting them uniquely from the fires that raged over the prairies each year. When the fires were stopped by the settlers, a rapid change took place in the oak openings: they became filled with dense strands of oak saplings. But surprisingly, the new oaks were not burs; they were largely blacks. Frequently a pure stand of black oak would spring up amid widely spaced bur oaks even though there might not be any mature blacks for miles around. Some early observers attributed this black oak irruption to a mass seeding of the openings by flocks of passenger pigeons. Each bird was presumed to have carried a single black oak acorn across the prairies to the bur oak grove and dropped it, in concert with his fellows. We now understand that no such charming an explanation is necessary. The black oaks had been there all the time, growing as brush or grubs among the prairie grasses, suppressed by annual fires. When the fires stopped, the black oak developed swiftly into tall, mature trees, gradually shading out many of the open-grown bur oak veterans. Today a prairie grove with an intact ground layer is the rarest plant community in the Midwest, yet there are more oak forests than there were in 1800.

I cite this bit of ecological history because it may help us to understand, I believe, what is happening, and what may happen, in the area of what has been called conservation information and education.

Since the 1900s the American conservation landscape has resembled a savanna. Here and there on a broad prairie of indifference you could identify scattered bur oak individuals, organizations, and agencies, their tough hides protecting them from annual fires of covert carelessness and overt retribution. Now, with breathtaking scope and velocity, the scene has changed. The oak openings of conservation today are thick with saplings. And these saplings are a different species, springing from submerged roots, displaying a distinctive foliage. This much we can see. What lies ahead is problematic. Will the black oak environmentalists grow straight and tall to form a dominant forest, shading out the bur oak veterans? It may indeed be that the militant ecology movement is leading squarely to a broad rearrangement of basic social and economic institutions, as John Pekkanen writes. Professor Wilson Clark, on the other hand, says environmentalism offers nothing really new, and that we had better get back to fundamentals. So, will the new cohorts succumb to renewed fires of convention, leaving only the old, gnarled sentinels of concern? Or could we see emerge a unique community of symbiotic relationships between old and new, exhibiting heterosis—hybrid vigor.

If those of us concerned with environmental communications and education are to continue to contribute effectively to the emergence of a broad ecological conscience, it will be helpful if we speculate sensibly about possible answers to such questions as I have posed. But before we can do so, we must ask several more. What characterizes the "new conservation"? Who is the "new conservationist?" What was responsible for his irruption? And then, finally, where are we all going?

The New Conservation

What are the differences between the old "conservation" and the new "environmentalism?" While some differences may be more apparent than real, others are quite distinctive. They may be summarized as follows:

In terms of its *scope*, the new environmentalism attempts to be all-encompassing. Whereas yesterday we tended to treat soil conservation, water conservation, forest conservation, wildlife conservation, and so on, as separate units, today we try to understand and explain the ecological unity of all man-land relationships. In terms of its *focus*, then, the new environmentalism is man-centered. That is, our primary concern has shifted from the survival of remnant redwoods and raptors to the survival of nothing less than the human species itself. At the same time we are not so much concerned about quantities of natural resources as we are about the quality of the human experience. "Conservation used to be merely a hobby practiced above the 10,000-foot level by a few eccentrics," James Bylin has written. "Today it has become a universal synonym for human survival."

In terms of its *locus*, while the old conservation conjured up images of open country, the new environmentalism incorporates the pressing problems of the city. In terms of its *emotional underpinnings*, the new environmentalism is based more on fear for man's tomorrow than on a love for nature's yesterday. Thus today's "preservationist" is not a lover of wilderness; he is one who fears the four horsemen of "conquest, slaughter, famine, and death." In terms of its *political alliances*, the old conservation was linked to such orthodox causes as depression pump-priming, national defense, and outdoor recreation; the new environmentalism, on the other hand, encompasses the hitherto unmentionable demands of the neo-Malthusians for population control.

It is in its *basic cultural orientation*, however, that the new environmentalism differs most strikingly from its antecedent, conservation. The latter, in the words of one patron saint, stood clearly for economic development, for the infinite goodness of American "progress." But environmentalism reflects a growing suspicion that bigger is not necessarily better, slower can be faster, and less can be more. As Saul Pett has written recently, "More and more people actively seek to conserve a tree, a lake, a view. More people question the Biblical injunction to be fruitful and multiply. More people question the old American faith in growth and expansion, and suggest that maybe what we don't need is another factory in town. More middle-aged people have begun to sense the validity in the young who scorn the plastic life." Upon viewing a new smokestack, millions of Americans used to see a sign of progress; now they see a sign of pollution. The mammoth motor car used to be a symbol of affluence; now it is a symbol of effluent.

If anything surely marks this revolutionary nature of both the rise and rationale of the new environmentalism, it would be the recent words of a Republican President of the United States, telling us that "wealth and happiness are not the same thing," that now is the time to "make our peace with nature," and that we must "measure success or failure by new criteria." Not even FDR would have said that!

If it all sounds suspiciously like echoes of Thoreau and Leopold, it only suggests that the black oaks were indeed here all the time, waiting only for a favorable concatenation of events to vault the philosophies of a Walden or Sand County muse into a State of the Union address.

The New Conservationist

Now, who is the "new conservationist?" He comes, of course, in pelage of many colors, and it will take years of sociological research before we can arrange him by phylum and genus. But on the basis of subjective analysis it seems to me we can take note of at least three types.

First, there is the old conservationist who has acquired an awareness of the global nature of what once seemed a parochial problem, an understanding of some new points of entrée toward constructive action, and a vastly heightened sense of urgency. He is the erstwhile County Conservation League member who has shifted his emphasis from prairie chickens to air pollution.

He is the Sierra Clubber who has added "human survival" to his agenda. He is the Park Service specialist who is trying to take his parks to the people instead of vice versa. Witness the words of the National Wildlife Federation: "Conservation is no longer just the story of vanishing wildlife and vanishing wilderness areas. There is a new urgency in the word today. Suddenly, as we stop and look at our total environment, it has taken on the meaning of human survival." Yet by no means have all old conservationists boarded wholeheartedly the ecological express. After all, like bur oaks, they have survived by resisting the fires of their surroundings, and they see the present situation as but another momentary diversion. We can expect to see them dotting the landscape indefinitely, lending perspective if not punch.

The second distinct type of new conservationist is a "she." But she is not the proverbial "little old lady in tennis shoes" who has graced the ranks of the bird watchers. She is the sharp young housewife whose automated kitchen has rendered her under-employed, and who, in looking around for new worlds to conquer, has discovered the environment and its problems. Through such local fire brigades as a Capital Community Citizens and such national organizations as the League of Women Voters she is lying down in front of bulldozers, accosting legislators in their lairs, baiting conservation bureaus, plumping for bond issues, and in general raising polite hell in the best traditions of American populism. She may even propagandize her Rotarian husband.

Perhaps the truest type of new conservationist is the committed college student who is making his presence felt through such activities as a national campus environmental teach-in. He is indeed a new breed in several respects. As a matter of fact, he is several breeds. At the far left is the true radical who sees environmental degradation as the Achilles heel of capitalism, and hence is riding conservation as his current hobby-horse toward revolution. For this "Mao-now" clique, the mouthy revolutionaries who profess to see some social Nirvana beckoning at the end of a trail of brutal confrontation, real conservation is the least of their concerns.

At the opposite end of the spectrum is the professional student in one of the resource management fields who is doing what comes naturally to him, in orthodox ways yet with early verve and élan. They are saying to their elders, in effect: "It is you and your system that have brought about the environmental mess which is making much of the world unlivable. Now, before it is too late, let us have a say about the profession and the planet that we are to inherit."

It is the students occupying the great middle ground of the movement who are unique. They represent everything from art to zoology. They come from Boswash and Grover's Corners. They have long and short hair, full pocketbooks and lean. Their folk heroes may be George Wallace or Bertrand Russell. They may know everything or nothing about the hydrologic cycle. Yet they have certain attributes in common: a neo-transcendental feeling for the man-land community, a revulsion for the excesses of a technological society, a suspicion of the Establishment, a sense of so little time, and a consuming desire to act now, and the devil take anybody who doesn't. As a Washington observer reports: "No group is more concerned, or more disgusted, about the growing destruction of the American environment than the young. First, they haven't been around long enough to become accomplices in the pollution violence, assuming they might want to. Second, the young are more concerned about saving the environment because they will be the worst casualties if it is not saved. Although many student environmental activists are using little more than the scream method, others are digging in for a long siege by finding out about the economics, the technology, and the politics of environmental problems."

The young eco-activist, in sum, is the same disenchanted American who came over in steerage, who pushed into the West. He is the same American who took off with a song for Bull Run, Belleau Woods, and Buna. Now the environment is his only frontier left, and the eco-war the only one he wants to fight. He insists on doing his own thing. He is willing to cooperate with the old-line conservation organizations, but he lives in fear of being coopted by them—that and the fear that a fatal public paralysis will render his cause impotent. He may call his mission impossible, but he still has an innate faith in the future, if only we act in time.

In essence, the practitioners of the old conservation have been exponents of the art of the possible. The new recruits see environmentalism as the science of the impossible.

The New Environment

Now, what has triggered today's mass irruption of the new conservationist? Historians, with the perspective of 50 years, have never been able to agree on the factors responsible for the so-called "first wave" of conservation at the turn of the century. Professor Samuel P. Hays says conservation had its origin in a concern for scientific management and efficiency among a relatively small group of planners and technicians. Historian Leonard Bates, on the other hand, argues that the movement was a grassroots upswelling of many passionate people versus the special interests of the day. From current experience, perhaps we can see that both of these explanations have their points.

Unquestionably today's environmentalism has its roots in the labors of a handful of leaders: particularly ecologists turned writers like Leopold, Carson, Darling, Commoner, Allen, Dasmann, and Ehrlich; and politicians turned ecologists like Udall, Nelson, Muskie, and Jackson. Unquestionably, too, the movement is taking on all the dimensions of a somewhat spontaneous general revolution in thinking, if not in action. In this transmutation of environmentalism from a learned cult to everybody's cause, many factors have surely been significant.

Professor John Gaus once pinpointed the critical elements in the ecology of any institution or movement as "people, place, physical technology, social technology, wishes and ideas, catastrophe, and personality." A brief examination of these factors at work in the America of the latter 1960s may illuminate the origins and development of our new ecological conscience.

The American *people* in the '70s are simply ready for the conservation message in a way they have never been before. They have been on a decade-long emotional binge that has left them both frustrated and pent up: multiple assassinations, civil rights confrontations, campus unrest, Vietnam, cost of living, crime in the streets—as F. Scott Fitzgerald once described a somewhat similar era, "all gods are dead, all wars fought, all faiths in man shaken." Little wonder that Americans are turning to their original font of solace, inspiration, and challenge; to nature, its ways and wise use. Countless individual leanings toward environmental concerns have been reinforced by the sense of community growing out of a timely Gallup poll, which indicated that conservation now is everybody's "thing." Another key social force is that evergrowing army of the young: Modern technological society postpones the age of work and responsibility. Many of the young must be trained through high school, university, graduate school, and apprenticeship. In the meantime, the student can afford the luxury of a strictly ethical view of the world—uncluttered by those compromises and deals that are the glue of any society. And in this long meantime, the student moralist is having a profound impact on politics—and on pollution abatement.

Yesterday's environmental degradation was usually over the hill and far away—in somebody else's dust bowl, somebody else's Echo Canyon, somebody else's boundary water canoe area, somebody else's forest. But the *place* of today's environmental degradation is where we live—in the foul air, rancid water, and clogged arteries of our cities. Millions can see, smell, taste, and hear the problem now: "The environment may well be the gut issue that can unify a polarized nation in the 1970s. It may also divide people who are appalled by the mess from those who have adapted to it. No one knows how many Americans have lost all feeling for nature and the quality of life. Even so, the issue now attracts young and old, farmers, city dwellers, and suburban housewives, scientists, industrialists, and blue-collar workers. They know pollution well. It is as close as the water tap, the car-clogged streets and junk-filled landscape—their country's visible decay," as *Time* reports.

Continuing along the Gaus outline, the *physical technology* of the 60s has been responsible for our current state of mind in a striking way. It has vaulted us to the moon, and thus has given us renewed faith in our capacity to conquer, but from our new vantage point in the cosmos we have also been struck as never before by the fragile, finite character of Spaceship Earth. By invading one frontier we have rediscovered another, the state of harmony between man and land. So we are appalled that the combined governmental expenditure at all levels on natural resources, including agriculture, amounted to less than $7 billion dollars last year, while we spent $9.7 billion for tobacco, $15.5 billion for liquor, and $5 billion for cosmetics. We are particularly appalled because the technological eye of the decade—the TV camera—has brought environmental degradation right into our livingrooms with stark realism. Along with scenes from Selma and Saigon, we have squirmed at scenes from Santa Barbara and Sanguine.

Developments in the *social technology* of the 60s have likewise played a part in the rise of eco-awareness and eco-action. The voice of the mass media has become increasingly dominated by a coterie of magazine editors and TV commentators, so when these gatekeepers of communications have seized on pollution as the big story, the snowball effect has been dramatic. Not in our wildest dreams as conservation people have we thought that *Time, Life, Newsweek, Look, Fortune,* and *Sports Illustrated* would ever devote simultaneous issues to ecology, yet that is exactly what happened earlier this year, accompanied by electronic voices like those of Walter Cronkite, Eric Sevareid, John Chancellor, David Brinkley, and Ed Newman, not to mention Arthur Godfrey, Eddie Albert, Johnny Carson, and Pete Seeger.

(What got the media onto the population-pollution story so compellingly? It is probably not without significance that the big media messages in this country originate from our two most populated and polluted places—New York and Los Angeles.)

The media calls to reveille might still have met with no response were there not throughout the land a spirit of rank-and-file activism, particularly on the part of the young. Whether you call this spirit a new "participatory democracy" or a throw-back to the Boston Tea Party, the result is the same—a confidence in the tactics of confrontation. Americans have witnessed the subtle yet sure effects of civil rights marches and peace moratoria, they have seen the results of brass-knuckles conservation as practiced by the Environmental Defense Fund, they have watched a David like Ralph Nader take on the Goliath auto industry, and

now they are ready to practice the same guerrilla warfare on a
broad scale for a cause that is easily identified with all the
better angels.

In their *wishes and ideas,* all the great ecological philosophers
have always expressed frankly the belief that true conservation
would require a profound change in American values. But nobody
really listened. We went right on basing our practices on economics at the expense of the aesthetic and the ethical. Now,
however, the youth of America are beginning to get the real
message that was there all the time. It fits right in with their
basic anti-materialism, anyway. Probably nothing so accounts
for the current popularity of conservation on our campuses as
this marriage of orthodox ecological doctrine with the innate
iconoclasm of the young. In environmentalism there is no generation gap: "It is possible that ecologists can eventually stir
enough people to an emotion as old as man—exaltation. Ecology,
the subversive science, enriches man's perceptions, his vision,
his concept of reality. In nature, many may find the model they
need to cherish," *Time* speculates.

Social scientists are saying that only once before in recent
American history has there been so profound and rapid a change
in American public opinion as the rise of eco-understanding.
That was the flip-flop from isolationism to interventionalism oc-
casioned by Pearl Harbor. So it is all the more striking that the
emergence of ecology as everybody's "bag" has not been triggered by a single *catastrophe.* We have had, in recent months,
our Everglades, our Storm Kings, and our Alaskas, but these have
not been continental disasters. Yet this is just the point. Today's
catastrophe is not a single clap of thunder, it is a pervasive driz-
zle, and thus all the more far-reaching. There is literally no dry
spot. Millions can sense that swelling population, rampant technology, and fragile biosphere are on a collision course, threatening the quality of the human experience if not the very survival
of man.

No single *personality* dominates the ecology of environmentalism as T.R. and F.D.R. dominated the first two waves of conservation. President Nixon has tried to preempt the movement,
but it is doubtful that he will be able to make his image stick.
Nor will any one scientist or interpreter likely run away with

the show. Rather, the new conservation is characterized by the diversity of its exponents and troopers. It is, in itself, a complex ecosystem, and this speaks well for its stability and longevity. Few of the conservationist troops know who their generals are, or even their sergeants. Perhaps one day a Bill Mauldin will capture in cartoons the personality of this new-style World War III, and it will not be a Patton, but the peace-time equivalent of G.I. Joe.

Out of the changing people, places, technology, aspirations, conflicts, and personages of America at the turn of the decade has come a new spirit and a new agenda. The spirit is an embryo ecological conscience. The agenda is clean air, clean water, open space, zero population increase—an illusive yet essential entente between modern man and the only home he has. In short, we sense that we have mortgaged the old homestead and nature is liable to foreclose. Or we may sense that the real spectre that pollution casts over man's future is not, perhaps, the extinction of *Homo sapiens* but his mutation into some human equivalent of the carp now lurking in Lake Erie's fetid depths, living off poison. Rene Dubos speculates that man can indeed adapt to almost anything, even the dirt, pollution, and noise of the city. And *that*, he says, would be the real tragedy, worse than extinction—a progressive degradation of the quality of the human animal.

Turmoil and Trends

Will we make it? What will our prairie grove look like 30 years from now? Will the black oaks of the new environmentalism be the dominant species, the old burs rotting and fracturing in the shade? It could happen. After all, the veteran resource management agencies and organizations are not particularly ecological in their orientations. On the contrary, they tend to espouse unilateral programs and cultivate special-interest clientele. For example, the Soil Conservation Service supports the drainage of the same wetlands the Bureau of Sport Fisheries and Wildlife seeks to preserve, the Corps of Engineers would inundate a national park without batting an eye, the Forest Service has never been

much of a custodian of wilderness, the Bureau of Outdoor Recreation already represents yesterday's patrician focus, and the
farm-subsidy program encourages the misuse of toxic chemicals,
one-crop farming that destroys ecological diversity, and mechan-
ization that drives jobless rural laborers into packed cities. If
these agencies and their publics are so inflexible as to resist coordination, they may well wither. Indeed, the white light of publicity we have all caused to fall on conservation can hasten the
demise of the organization that temporizes in adjusting to changing times, changing priorities, and changing audiences. As Dean
Kenneth Pitzer says, "We can make no greater mistake than to
shrug off the ideas of the young as foolish and impractical. It
will be at our peril that we encourage them to point the way to
a better world, only to tell them later that nothing can be done
about it. Conversely, youthful idealism and energy represent a
potent force for good if only we can give it productive and cre-
ative outlets. We must be willing to change."

On the other hand, our prairie grove of the year 2000 may be
punctuated only by the old bur oaks. It could happen. The new
environmentalism could turn out to be only a passing fad, like
hula-hoops. Americans are given to switching issues in mid-
stream. Whatever happened to United Nations Day? Some are
already saying that while "the conservation and proper use of
natural resources may be a fundamental problem, that should
not divert the nation's attention from the problems of social
justice and racial equality." Writes Bob Beyers, "I particularly
note a growing resentment among blacks that environmental
interests may represent a white 'cop out' from pressing issues of
race and poverty."

Or perhaps the sonorous voices of gloom and doom will render
us absolutely insensitive to any possibilities for progress. The
fallout scare of the 50s produced precious few underground
shelters. You already hear the complaint: "We have read the
statistics of degradation, and heard them, and flinched at them,
and even wearied of them; statistics that boggle the mind, and
that are repeated like clockwork every year, inching higher and
higher. It takes something really different—like a river so filthy
it actually catches fire—to engage our jaded attention," according to educator James Allen. The Jeremiahs of the movement

could indeed have such a narcotic effect on public opinion that masses of Americans will "tune out." Alvin Toffler has already written of the danger of "future shock," which forces us to crawl into our shells in utter despair.

The new recruits may also just plain run out of gas "when the bandwagon stops coasting and has to be dragged up the hills." Or they may climb off "when the limits of present ecological expertise become apparent."

There is another danger: If the eco-activists on college campuses become dominated by neo-Fascist hooligans, irreparable damage will be done to the new environmentalism. Riots will repel, not attract, support. The single thing more dangerous to man than environmental pollution could be the growing clamor over the issue, according to Robert Theobald. Intense efforts to change the established social order could produce a reactionary backlash that would pit man against man in a most unecological fashion.

Hopefully there can be a third broad possibility—a wedding of old and new. In many cases today, what was once a natural prairie grove is dominated neither by bur oaks or blacks but by thrifty white oaks. These white oaks exhibit some resistance to fire as well as a tolerance of shade. Perhaps we will likewise see emerge a lay ecologist with all the lore and savvy of the old-line conservationist and all the heightened idealism and sense of mission of the new environmentalist. We need both, as reporter John Margolis has pointed out. If the traditional conservation organizations had just spent all their time worrying about ecology, fewer woods and waters would have been saved in recent years. If the new conservation is to succeed it will be through the mainstream of going groups, he says, yet groups that grasp the technical and radical rules of the new ballgame. We must seek a conservation movement old enough to have traditions and young enough to transcend them.

In this regard it is particularly intriguing to see the hesitant yet healthy emergence of an eco-industry. We are beginning to hear of "ecological stocks" pacing the New York exchange. If Americans can learn to make as much money out of environmental husbandry as we have out of environmental exploitation, the problem will be solved with some dispatch. It will be particularly important that the private sector lend a massive hand, because environmental reclamation is not a nice, neat governmental package like a Manhattan project or a lunar landing. It is more like the Depression or World War II in its breadth and diversity. True, attacking environmental ailments has a special appeal for Americans; in large part they are technical and mechanistic problems that involve processes, flows, things, and the American genius seems to run that way. So a typical American response to a series of crises in a given field has been to smother it with money and expect solutions to appear promptly. "But environmental-quality problems do not lend themselves to this kind of approach. There is no single goal toward which technology, economics, and management can marshal their forces. Environmental pollution and degradation appear in many forms and many places, and successful programs of prevention and amelioration will be difficult and many-sided. The complex scientific-engineering-economic-political-management-educational programs for cleaning up the air, the water, and the landscape will have to be tailored to meet different situations in various regions of the country, in various industries, and in various social conditions," according to economist Joseph L. Fisher. Furthermore, practical ways to resolve the obvious conflict of economy and environment are far from clear for a free society. There is real danger that the current emphasis on the importance of ecological balance will obscure the importance of economic balance. Certainly we cannot go back to some Medieval womb. We must start with what we have and work forward.

One way to make the economic system accountable for the damage it does to the environment is to work the costs of avoiding environmental damage right into the pricing system. As higher prices then begin to show up for goods whose manufacture damages the environment, the mass re-examination of values necessary to the beginnings of broad environmentalism may begin to take hold. In a real sense the revolution in the way we view things may already be happening. That is, we may already be seeing a mutation of bur and black oaks, so to speak. For a particularly perceptive analysis of this new "counter-culture," I quote from Edward Kern in a recent issue of *Life:*

Gradually but imperceptibly, the ties that held people to the old ways of thinking are loosening and new ties are being formed to a new outlook. It is often said that we are in the midst of a social revolution. The truth, probably, is both something less and something considerably more. There is a social revolution, which seems only to have begun; but there is also something more profound—a revolution of consciousness. Conceivably it could alter the whole aspect of America and produce a new species of American. If it does, this would not necessarily doom the existing structure of institutions or the present forms of political life. The impersonal pressures of advancing technology are certain to have a great effect on these, and trends point to larger and more complex organizations in the future rather than to the simplicities of the counter-cultural commune. But organizations, from one standpoint, are social vessels which are designed to contain humanity. What matters most is the quality of human consciousness that is poured into them, and time may prove that it is possible, after all, to pour new wine into old bottles.

The Big Test

In summary, what is really on trial in the tension zone of the '70s is not a movement. It is the processes of American education, American democracy, and American culture.

Can education go beyond a mere imparting of ecological facts to inculcate a will and a way for the individual and collective solution of environmental problems?

One difficulty in changing attitudes toward environmental exploitation is that attitudes toward the environment are tied to over-riding values that are highly resistant to change. There are many basic values in Western culture that support environmental destruction—the Abrahamic concept of land, for example. But there are also basic values that support environmental conservation—the survival instinct, for instance. We need not be discouraged. When threatened, man is capable of almost anything. Today, nothing less than our survival is at stake. The problem is getting enough people to realize this blunt truth while there is still time to act. Yet fear will not be enough as a longterm motivating factor. We need love: "The one thing needed to recover and preserve the American environment is a reverence for earth—paying fair homage to the soil, the winds, the waters, and honoring the very spirit of their places," says Coleman McCarthy. Whatever the theme, the practitioners of conservation information and education have one easy test of their solvency today: If they are not drawing fire from the flanks, what they are putting out is not getting to the heart of the problem. There is simply no easy compromise between the old economics and the new ecology. All this suggests to us that we must "get with" producing an enlightened citizenry that will, in the words of the author of the Environmental Quality Education Act, understand "man's unquestioned interdependence with nature," appreciate that "scientific advance is not always synonymous with progress," and "use an ecologic filter when making important policy decisions."

Can industry and government move beyond mere tokenism in their response to environmental degradation with imaginative programs that demonstrate the relevancy of capitalism and democracy to gut issues?

The political and economic conflicts growing out of militant ecology will be enormous, writer Luther Gerlach predicts: "The question is whether or not those responsible for damaging the environment will be wise enough and adaptable enough to see what is being demanded of them and accommodate to it. The key to holding us together is how business and political leadership responds. Confrontation is unavoidable because the environmental problem does not lend itself to tokenism—too many people are aware; too much is seen; the crisis is too great. The strongest argument for optimism is that the leaders of the movement are the educated people who know where the levers of power are and who are willing to use them short of taking to the streets."

Can our society create a new consumerism that demands less goods and gadgets, and more capacity to preserve, protect, and defend our natural heritage?

The answers to most population-pollution problems can only be found in trade-offs. If something undesirable is to go, something desirable may have to go as well. You can't air-condition your home, for example, unless somebody is burning fossil or nuclear fuel to produce electricity. It is the making of sophisticated choices, then, the rendering of subtle value judgements that is the essence of conservation today. Our first task is a good old American goal—to restore more freedom of choice. The consumer who *wants* to conserve must be given the *chance* to conserve—in the marketplace, in his home, at the ballot-box. And then we in positions of leadership will be increasingly put to the test of outlining the options in an unemotional, objective, self-disciplined manner.

The United States simply must start inventing genuine "political, economic, and intellectual processes that will give us, as a society and as individuals, more real choice about how we live." These are not the words of a Berkeley fanatic; they are the words of *Fortune* magazine. We must, in short, begin to engage, once and for all, in ecological thinking. As someone has said, we cannot ever do only *one* thing. When we try to pick out anything by itself, we find it inexorably connected to something else. And we cannot do everything all at once. We must rid ourselves of the mentality of the 30-second commercial and the 30-minute comedy. Environmental housekeeping is a never-ending serial.

Put another way, what is really at stake in the '70s is us. We must demand of ourselves the same high quality we demand of the environment.

What is necessary is an unflagging respect for the world, and for man. For dissent and diversity. For those natural amenities that husband the prosperity of the human spirit. For those human institutions that protect freedom of choice. If we simply regret what we have done, we must regret that we are men. It is only by accepting ourselves for what we are, the worst of us and the best of us, that we can hold any hope for the future.

To paraphrase educator John Gardner:

We will not find our way out of our present troubles until a large number of Americans—and outdoorsmen particularly—acknowledge their own special contribution. The path to recovery will call for courage and stamina. Our salvation will not be handed to us. If we are lucky we will have a chance to earn it.

Many things are wrong. Many things must be done. There is no middle state for the spirit. It rises to high levels of discipline and decency and purpose—or it sags and rots. We must call for the best or live with the worst. This is everybody's ecology in the environmental decade.

Index